Beyond Disease and Dissection: A Holistic View of Brain Function

Jemison

Contents

Introduction

Contemporary neuroscience is an ambivalent discipline. There are large efforts in numerous subfields, but, on the other hand, it is difficult to obtain a coherent overall picture and a general understanding of the nervous system and of the brain from all these investigations. Neuroscientific research spends major efforts on special aspects of the nervous system and on detailed phenomena, performance features, brain areas or functional circuits. Data is gained from neural defects or diseases – as, e.g., Parkinson, Huntington, addictions or lesions of special brain areas –, the behaviours of bored captive animals are investigated – as, e.g., Macaques or rodents –, this is aligned with anatomical findings and magnetic resonance images, and interpretations of special phenomena are issued – as, e.g., perception, pain, emotion, memory, reflexes, motor activity, language, attention, plasticity, cognitive capabilities, consciousness and social aspects. Lots of information and findings are interpreted in segregated manner by this way, but this is typically examined in multifarious studies with few regard to each other or to external aspects, known phenomena or to research results from other part disciplines. This leads to a picture to the research results, which is greatly multifaceted, but poor in its comprehensive understanding of the matter.

I am pretty convinced that the nervous system is indeed a "system", which functions as a coherent whole for any purpose. From my point of view, the neurosciences and each of its part disciplines should always spend significant efforts to support this view and to explain the brain as a unified system. But this does not occur to a serious extent, at least not as far as I have been able to take note of the concepts. That's why I am disappointed with the contemporary approaches, research strategies and results in this discipline and why I am looking for a way out of this shortage.

A special example of the misery in neurosciences is the usual dealing with the concept "self-regulation". Wagner and Heatherton (2014) investigate the capacity of humans and animals to "inhibit prepotent responses in order to obtain later, larger rewards (i.e. delay of gratification)" (709). They explain self-regulation as general capability to control behaviour: "At its core, self-regulation is concerned with starting, stopping, or modifying thoughts, emotions, or behaviour in order to pursue goals or stay in line with societal norms." (710) This is what I would actually associate with the concepts "self-control" and "impulse control", but never with "self-regulation". Yes, self-regulation should include this type of self-control, but it should actually include much more aspects and boundary conditions. I would tend to reserve term "self-regulation" for a matter which is much more complex, and which refers to the capability of the human to regulate his affairs under real-life conditions, comprising evolutionary pressure and stress due to competing needs, competing natural laws, competing fellow man as well as due to competing environmental and social developments. Self-regulation is a term, which cannot be missed, when trying to explain the whole complexity of human existence, but Wagner and Heatherton (2014) tend to block this concept for simple self-control under no (remarkable) circumstances or under a specialised test condition – that the choice between two rewards must be made.

The dealing with the concept "reward system" is the next shortcoming. It is a usual way to investigate behaviours and neural processes in scenarios with rewards with different values and probabilities – see e.g. Rushworth et al. 2014, 504. This is just one of the few abilities to bring investigations on several types of neural processes, as, in this case, of the decision making system in the frontal cortex, a step forward. But this type of research seems at the same time to induce a perception on the reward system, which is totally misleading – it draws a picture at which rewards are seen as an idiosyncratic quirk and a weakness of humans and animals. But I would tend to see the reward system as one of the most important features of the human brain, which is the actual basis for the development of culture and economic growth. Of course, reward related behaviours may sometimes turn into weaknesses, but this is only a side effect. Actually they are one of the greatest inventions of the Evolution, what will be discussed in this book in connection with the concepts "artificial/virtual needs" and "growth needs". But neural researchers, as e.g. Rushworth et al. 2014, tend rather to describe the reward system as a special peculiarity of mammals and humans, without mentioning that this might be one of the greatest inventions on the way to the contemporary society.

Another point is the relationship between information and purpose. It is clear that it is one aspect of the brain that it is an information processing engine, which exploits perception for produce controlled behaviour. But it is also clear, that it is no computer. Contrary to computers, it is induced by purposes and inherent regulation mechanisms, which are incorporated as motivations and emotions. This is also information, but which plays a specific role and which causes continuous self-optimisation. Contemporary neuroscientific concepts do regularly not indicate this aspect thoroughly, or they ignore or deny it completely (see, e.g., Tononi 2012 and section "Tononi, G.: The integrated information theory of consciousness" in chapter "Consciousness" in Part 2 of the book).

That concepts are developed into wrong directions, which lead into dead ends rather than to a coherent understanding of the whole nervous system as well as of the system *hu-*

man under real-life conditions, is symptomatic for the neuro-sciences. I would not suspect that these aberrations are made by intention, no! But they might be provoked by the limitations of the research methods, which are available today, and by the extremely specialised research culture.

Functional magnetic resonance imaging (fMRI), a method which is an important basis for contemporary neural research, is just not really functional, because it visualises only metabolic processes (power supply for information transport) rather than neural excitation processes (information transport; note: it will never be possible to understand a computer by concentrating on the currents in its power supply units). And even if functional magnetic resonance images (fMRI) would be *functional*, it is just not possible to make such images of managers or alpinists under real-life conditions. And even if this would be possible, it might hardly be possible to interpret the complex neural data and interrelate it with the complex contexts of action and cognitive operation in these real-life scenarios. That's why neural research must be specialised and can only investigate particular phenomena under simplified boundary conditions. So it is no wonder that it produces large amounts of ambiguous findings rather than general models. But it would be recommendable to cultivate also always the latter aspect.

It is important to get a comprehensive understanding of the whole system, reaching from the most detailed micro structures and atomic processes, i.e. from the micro level, up to the general conceptual ideas, i.e. the macro level. Because of the highly specialised research in this day and age, there is actually no lack of micro-level models. But the macro level is largely underrepresented, not least because of the extreme complexity of the nervous system apparatus. This book tries to make a contribution in this field, by proposing a special macro-level model, called "suitability probability processor model" (SPP model). I am sure that there are much more ingenious solutions implemented in the nervous system than nowadays known, namely at the macro level. The SPP model is an attempt to discover some of these ingenious concepts.

The *suitability probability processor model* (SPP model), which is proposed in this book as an appropriate macro model of the brain, is just an attempt to capture a more systemic claim than available until now. Special advantages of this model are that it involves rewards, purposes, emotions, motivations as well as self-regulation and self-optimisation under real-life conditions as crucial components, and that it tries to combine all important aspects to a coherent concept. But this can only be a small step in the right direction, if at all, while remarkable progress might need big efforts of generations of scientists at both the macro level and the micro level.

It is difficult to achieve progress in understanding the brain as a coherent system, and the SPP model does not contribute additional empirical findings – it is rather only another macro-level interpretation of well-known facts and part models. Thus, it is questionable if such a general model can be a valuable contribution on the path of knowledge. But, on the other side, it would be a failure to neglect such kinds of attempts, at least if there are ideas available of how to solve the brain puzzle in an unconventional way.

In this book, the brain is characterised as *suitability probability processor* (SPP) in contrast to computers, which are arithmetic logic processors. Both types of engines process information continuously, at which subsequent operations are dependent on precedent operations or external events. How the sequence of operations or the reaction to external events is selected in each case, depends on logical or binary decisions, respectively, on computers, and the main difference of brains, compared to computers, is, that brains follow a completely different paradigm in this respect. They base sequences of operations as well as reactions to external events on a suitability probability evaluation process, which is continuously ongoing in crucial brain areas. This is a central book of this book, and it is the reason why concepts like "suitability probability evaluation" and "suitability prob-ability processor" are introduced here. What this means is explained step by step in chapter "The SPP model". Note: There are many differences between brain and computer, but there are also commonalities on an abstract level – more de-tails see in section "Commonalities and differences between brain and computer" in the second-last chapter (in Part 2).

The book is divided into two parts: Part 1 – "The Brain as Suitability Probability Processor" and Part 2 – "Excursions to the current state of science". Part 1 provides the complete description of the SPP model and of some supposed con-sequences. Part 2 comprises additional background informa-tion, composed of references to important scientific sources and further discussions of special aspects. It is recommend-ed to read Part 1 and follow the numerous references to Part 2, which are included in Part 1, rather than to read Part 2 continuously.

Before we start, it should be made clear that the human is composed of at least 3 complex systems and that we will pri-marily focus on one of these systems. The three systems are the body with its biochemical (physiological) structures and processes, the central nervous system (CNS) with the brain, which regulates human's affairs in his environment, and the combination of autonomic nervous system (ANS) and neu-roendocrine system which takes control of the well-being of the body by mediating between body and CNS. Our focus lies on the brain and the

body, ANS and neuroendocrine system are only

tem (PNS), which connects the CNS with limbs and body, and which includes the spinal cord and peripheral nerve cords; this is seen here as being subsumed under CNS; in the functional sense the PNS is indeed not a separate system.

There are many introductions into the brain structures and processes available. So it is rather not necessary to add another or better one. But we will nevertheless start with a short introduction into some "neuro basics", because it is necessary to prepare the ground for the description of the *suitability probability processor model* (SPP model), which will follow.

Neuro basics

Note: This chapter summarises basic knowledge from various popular sources (without known references) as well as information from Popper and Eccles 2006 and Gazzanega and Mangun 2014. Beside this, it is also influenced by findings and theses that are proposed or discussed in this book, particularly in Part 2, but also in the other chapters of Part 1.

Purpose, perception and motor control

The task of the central nervous system (CNS) is basically very simple: It has to fulfil a purpose by controlling appropriate actions on the basis of suitably filtered information from sensory perception. This is and remains the ultimate principle, independent from the complexity of the implications which may evolve over time.

The **purpose** is survival and thriving as social being under the conditions of evolutionary competition. This has to do with metabolic homeostasis, prevention of painful states, satisfaction of needs, emotions and rewards. Stimuli and control loops of these types are incorporated in the brain processes via multifarious mechanisms. This matter refers to concepts like hypothalamus, endocrine system, homeostatic regulation, pain sensation, limbic system and reward system. This is summarised hereinafter as "emotional-motivational system" (see also sections "Homeostasis, pain, emotions and rewards" and "The emoti(onal-moti)vational system").

Perceptual information is gathered by the five classical senses – vision, hearing, taste, smell, touch – and also by a couple of further senses, like temperature, as a variant of taste, balance, pain, proprioception etc. This means always that sensory receptors or organs stimulate afferent nerve cords, which send electrochemical signals to synapses in the spinal cord, subcortical circuits or sensory areas in the cerebral cortex. These sensory signals elicit typically as patterns of multiple stimuli and they are transmitted via multiple nerve cells to the target brain areas and target synapses. They take effect in various ways by different types of information processing, which might be called transformation, projection, communication, distribution or funnelling. The universal basis for this informational processes is the organization of the brain as a network of nerve cords which are linked over an almost endless number of synapses, but which are typically established and adapted in a well-organized manner. The two basic transport mechanisms at synapses are excitation and inhibition.

Appropriate **actions** are produced via specialized areas in the so called "motor cortex", electrochemical transmissions of signals through efferent nerve cords to muscles, and muscle contractions which are caused by this way in a controlled manner. These processes can only function sufficiently if occurring well-integrated with other types of processes, as, e.g., sensation (see above) – mainly also via the so called "somatosensory system" (proprioception) – and regulation of vigour, balance and smoothness, e.g. via circuits in the cerebellum and basal ganglia.

One of the most important features of the brain is also that perceptual and sensory information is not processed in its entirety, just as it occurs in environment and body, and as it excites the afferent nerve cords in every second, but that this huge bulk of information is reduced to the relevant subsets by clever filter and transformation processes. This leads to the basic structures and information processing principles of the brain.

Excitation, inhibition, pattern transformation and circuits

Smart information processing is not only required for filtering and transformation of sensory information, but also for all other tasks of the brain, as, e.g., activity control, pattern recognition, evaluation of present and memorized patterns, and decision making. The basic solutions for these types of tasks, that might have to be exploited in very complex combinations and scenarios in the end, are again quite simple.

The brain, particularly the part which is called cortex or neocortex, is organized in modules, each consisting of a network of synapses between greater amounts of neurons. A synapse is a link between one neuron and another. It can influence the degree of excitation in the target neuron on the basis of the degree of excitation in the source neuron. A synapse can either have excitatory character, so that the excitation in the source neuron increases or amplifies the degree of excitation in the target neuron, or it can have inhibitory character, where excitation in the source neuron decreases the degree of excitation in the target neuron. That also inhibition is available as counterpart of excitation is essential for the excitation waves which flow through the brain. Without this feature, the neural networks would not be able to balance any process, and there would be nothing else than constant excitation overflow.

"The yin and yang of excitation and inhibition probably arose as an early feature in the evolution of the nervous system. In his studies of spinal cord reflexes, Charles Sherrington found that excitatory and inhibitory neurons were always in tandem and that together they provided the algebra of the nervous system: 'The net change which results there when the two areas are stimulated concurrently is an algebraic sum of the plus and minus effects producible separately by stimulating singly the two antagonistic nerves' (Sherrington, 1908, 578). The algebra of the cortex has proved to be more elaborate, but the principle of controlling the balance between excitation and inhibition is a fundamental property needed for cortical computation, and not simply as a means of stabilizing excitation." (Douglas and Martin 2014, 81)

The direction of the information transport over synapses has typically one-way character. Each synapse has a certain transmission factor, which is basically stable, but which can be justified. The plasticity of the brain is based on the ability to establish and dissolve synapses and to justify the transmission factor of each synapse in fine gradations.

But what are the findings for a brain module. How is information processing, so processing of excitation patterns and flows, organized in a module? This can be made clear by the statement that modules are typically organized as multiple layers of neurons (typically six layers) in which each layer has multiple synaptic links to the next layer. So a module is typically a synaptic multilayer network. The way they are organized is partially described in the literature as maps or spatial maps: "We know that most classes of information represented in the cerebral cortex are topographically encoded on anatomically two-dimensional 'maps'. The cortex is made up of dozens of these small topographic maps. The two-dimensional structure of the primary visual cortex, for example, provides a topographic map of the world as seen by the retina [...]. At each point on that map, the firing rates of neurons tell us something about what the retina sees at that location in the visual world. What is hugely important is that this basic organizational structure has turned out to be almost entirely universal in the mammalian brain." (Glimcher 2014, 684, and references therein)

Explanation: "firing rate" means excitation, i.e. excitation is transmitted in nerve cells and over synapses as electro-chemical impulses with greater or smaller firing rates.

It is known – from neurosciences just as from informatics (artificial intelligence) – that neural multilayer networks are excellent performers of pattern transformation tasks. This means that, for instance, the following types of tasks can easily be solved by a well-trained synaptic multilayer network; well-trained means that the transmission factor of each synapse is justified precisely for the task of the network:

- Projection, transformation: A picture is provided as pixel array, by a neuron per each pixel. The excitations of the neurons represent the brightness of the pixels. The picture contains an object (a house) which can be larger or smaller, dependent on the distance. Scale the object to a uniform size, independent from the size in the original neuronal excitation pattern, representing a pixel pattern.
- Object detection: An array of neurons represents a picture, either sent by the retina or by another transformation module, as e.g. the network from the example above. The picture contains an object, e.g. a rectangle or a house or a cat. Detect the type of the object.
- Decision making: A large collection of input neurons represents information that has been transformed by upstream modules from various primary sources, including external sensory information (sight, hearing), somatosensory information (position of body, arms, legs etc.) and stimuli of pain, which are incorporated via pain sensation. A decision has to be made, what is the right response in terms of an appropriate activity procedure, to be elicited in the motor cortex. An appropriate multilayer network works as decision making map for the final selection of the winner pattern in this case.

The brain is able to process at each second a multitude of such types of tasks in parallel. This is only possible with complex projection cascades, what means that the output information from one module is typically transmitted or distributed to further target modules and that this occurs again and again over several cascades until the right information processing results can be achieved. But it is clear that such a strategy can only work for a system which is equipped with effective adaptation abilities and with a mechanism that forces continuous self-optimisation. But the brain is such a system. It has been optimised by evolutionary forces over ages.

The result is a sophisticated structure, which is described by Douglas and Martin 2014, "Organizing Principles of Cortical Circuits and Their Function", and references therein, as follows (79):

"A quick look at the composition of the neocortex is instructive. In each cubic millimeter of cortical gray matter we find up to 100,000 neurons, about 100 million synapses, and 4 km of axon [...]. By contrast the white matter, which is generally thought to contain most of the wiring of the brain, contains only 9 m of axon per cubic millimeter. These bald numbers suggest that the processing strategy of the neocortex is to exploit direct broadband transmission for local circuits in the gray matter and to use the white matter only for long-distance, low-bandwidth transmission. The implication of this is that most of the synapses made in any area originate from neurons within that area, not from outside sources, like the thalamus or other cortical areas. Thus, we should first look to the local circuits if we are to understand how the neocortex performs its multifarious tasks of perception, memory, cognition, and action. Of course, the long-distance external inputs to these local circuits are not incidental players, since some of them connect cortex to the peripheral structures, like the spinal cord, while others connect cortex to itself and to other essential machinery like the basal ganglia, claustrum, brain stem, and cerebellar nuclei."

Explanations: The neocortex is the largest part of the cerebral cortex, attributed to sensory perception, motor control, cognition, reasoning and language. An axon is an extension of a neuron (nerve cell) for the transmission of excitations to other neurons or to glands or muscles. Thalamus, basal ganglia, claustrum, brain stem and cerebellum are subcortical brain areas with prominent control functions.

Another point is that connections between grey matter modules via white matter pathways are not only organised in one-way directions or cascades, but that there are also loops and circuits. This means that excitations might flow from one module in a special brain area to another module in another brain area, that it might there take excitatory or inhibitory effects, and that the resulting signals might turn back to the original module in any form. This kind of loop organisation or circuitry might also realize circular interconnections over multiple modules and brain areas, enabling mutual interaction and feedback control, wherever identified as beneficial by evolutionary optimisation. This basic capability is particularly exploited for the interactions between the cerebral cortex (e.g. motor cortex or frontal cortex) and subcortical areas (e.g. basal ganglia or hypothalamus), but also for any other interactions and control loops (e.g. between the different language areas).

The division of the brain into areas and grey matter modules, and the installation of appropriate white matter pathways, control loops and circuits in a sensible way should be the crucial key to its effectiveness as a unified system. It is hardly imaginable that a human designer is able to construct such a brain apparatus to a sufficient extent, even if he would have comprehensive knowledge on all principles, what he preliminary does not. But the evolution, as a designer who is much more awesome than the human, was able to do this; though only on the basis of a resource which seems rare to us: time.

This was a short glimpse of certain basic organization principles of the brain. But important features have not been examined yet, like memory, associations, emotions and needs – this is the topic of the next two sections. And nothing is told about the control mechanisms and the question how the brain works and optimises as a whole and how this is driven by the demands of environment, body and human needs – this is attempted to be answered by the SPP model in the next chapter.

Memory

Memory is primarily divided into procedural and declarative memory, at which only the contents of the latter can be consciously remembered.

Procedural memory, so the non-conscious part, refers to the sensory areas and motor areas of the cortex which are responsible for perception and motor control. This is the part of the cortex where all kinds of well-trained reactions and behaviours are stored. This includes processing of perceptual patterns via cascades of sensory modules, which are well-adapted to the behavioural purposes. Each kind of motor activity can only be executed in controlled manner if guided by rich amounts of sensory information, what involves typically multiple senses. The procedural memory is the location where appropriate skills are stored in terms of sensorimotor procedures.

Although this memory content cannot be consciously remembered in full detail, it can nevertheless indirectly be included into memorizing processes and it can always be (re-) activated. This works via the declarative memory, as we will see.

Declarative memory divides into associative long-term memory, episodic memory and short-term memory. The **association cortex** is the location where the external world and the own existence in this world is internalised. This takes place via connections between sensory experiences and motoric capabilities, that are coded in secondary, tertiary and further specific (e.g. language) and unspecific associative areas in the cerebral cortex. One can, e.g., feel an object by cutaneous sensing, see it by visual detection and name it by linguistic processing, with the result that this object (or this type of object) is represented in the involved parts of the associative networks. This leads to an internal world view which is constantly modified and developed and where no piece of memory is stored without integration into the existing interconnections.

The access to this store, writing and reading, leads over the short-term memory (see in the following) and the control processes of the brain (see chapter "The SPP model" further down).

The **episodic memory** is located in cortical areas that are called hippocampus and temporal lobe. These regions are known for capabilities like spatial navigation, perception of time, episodic memory over weeks and mediation between the brain control system (or short term memory) and the associative long term memory.

There are two further remarkable characteristics – integration of procedural memory and of emotional-motivational sensation. All **procedural sensorimotor excitation sequences** that are involved in real-time activity, are always also part of the traces in the episodic and associative memory. This occurs in the form of microrepresentations of the procedural patterns. These can henceforth be used to include these opportunities into memorising and evaluation processes and to reinstate the detailed procedural patterns whenever necessary (see also Part 2 – chapter "Emotion, motivation and memory" – section "Microrepresentations of sensory-perceptual experiences, actions, internal thoughts, and emotions in the episodic memory"; see also Davachi and Preston 2014).

Emotional-motivational sensation is bound into hippocampus and temporal lobe via specific pathways and circuits. This has mainly two important consequences: These signals are integrated into each piece of memory in terms of emotions or feelings, and they take control over what is stored with which strength in terms of positive or negative values, to be encoded as excitatory or inhibitory effects at synapses. More details about emotional-motivational sensation see in the next section below and at the beginning of the next chapter.

Short-term memory does not exist as a dedicated store. What was meant by this concept has rather to be attributed to the control processes that are constantly ongoing in the brain. This is examined below in section "The control levels of the central nervous system" in chapter "The SPP model".

Notes:
- The statements from the last two paragraphs above might be scientifically controversial. They are close to the border between the current state of science and the theses that are proposed in this book.
- The concept "working memory", which is also sometimes used, can be interpreted in various ways – this discussion will not be adopted here.
- For more details about memory see chapter "Emotion, motivation and memory" in Part 2.

Homeostasis, pain, emotions and rewards

As stated above, the central nervous system of the human is devoted to the purpose of his survival and thriving as social being under the conditions of evolutionary competition. Perception, sensorimotor functions, memory and other "informational-functional parts" of the nervous system provide the capability to comply this requirement. But a crucial question is, how the necessity to follow the evolutionary pressure is bound into the brain processes – this is essentially realised via the following four systems: the autonomic interface, involving homeostatic regulation, the sensory system, namely pain sensation, emotional processing, and the reward system.

1) The **autonomic interface** integrates the nervous system with the autonomic nervous system and the metabolic processes of the body. The main interfaces for these processes are a subcortical area which is called hypothalamus and the endocrine system, including a larger collection of glands, like pituitary gland, pineal gland, thyroid gland, adrenal gland, pancreas gland, ovaries gland, testes gland and others.

The core part of the control processes which occur on the behalf of these interfaces is called **homeostasis**. This comprises, according to Maslow 1987, 15, and references therein, the regulation of metabolic balances like "(1) the water content of the blood, (2) salt content, (3) sugar content, (4) protein content, (5) fat content, (6) calcium content, (7) oxygen content, (8) constant hydrogen-ion level (acid-base balance), and (9) constant temperature of the blood [...] minerals, the hormones, vitamins, and so on". This is linked to feelings like hunger, thirst, cold, hot, sexual arousal etc. This includes also a control loop that is called hypothalamic-pituitary-adrenal (HPA) axis and which is connected with the regulation of stress levels in the nervous system and the metabolic system in consistent manner.

This includes also regulation of the balance between sympathetic and parasympathetic nervous system, which are two parts of the autonomic nervous system. The former represents preparedness for action and the latter rest and digestion, and the autonomic regulation processes include always regulation of the balance between both aspects of life, and thus between sympathetic and parasympathetic nervous system.

2) The **sensory system** has two main aspects:

- Information: provide signals for sensorimotor control or for the informational aspect of cognitive processes and decision making processes (see something, smell something or feel something which is not painful).
- Motivation: **pain sensation**.

Only the latter aspect is of interest here. This refers primarily to the sense of touch and the somatosensory system, whose signals are typically bound into cortical processes via a subcortical area which is called thalamus. This comprises sensory signals representing, for example, touch, vibration, heat, (the somatosensory component of) balance and pain. However, only pain is of interest here. This takes partially effect via explicit pain receptors in body and skin, and partially via the normal receptors if they are confronted with overmodulation, e.g. due to too much pressure or heat.

Besides this there are signals of pain which are produced by other senses, like vision (dazzling light), hearing (loud noise) or taste (bitter). But these are difficult cases – the transition between information and pain is widely adaptable, and domain specific investigation is needed for understand how these types of perception patterns turn into pain and motivational stimuli.

3) **Emotional processing** is particularly attributed to a

brain area which is called amygdala. This is associated with extraordinary excitation states that are caused by alarming environmental stimuli. This comprises fear and anxiety, but also pleasure and delight and a wide range of other feelings, which occur due to interactions with the natural and social environment. The amygdala, and thus emotional processing, is closely linked to other areas and systems, like habenula, which represents the olfactory sense, the autonomic interface or the reward system (see next paragraph).

4) The **reward system** is connected with concepts like mesolimbic dopamine system or mesocorticolimbic dopamine system. It represents emotions and the pursuit of any type of good feelings or rewards, at which this is seen as an open system with no determination or restriction, what kind of desires might occur with this behalf. The reward system is mainly represented by the following subcortical components: amygdala (representing emotions), septal nuclei, nucleus accumbens, ventral tegmental area and pathways to several substructures of the basal ganglia (striatum, caudate nucleus, putamen, substantia nigra etc.). Neurotransmitters, which play an important role in this system, are dopamine, serotonin, glutamate and gamma-aminobutyric acid (GABA).

This was a short overview of the four major systems which convey the evolutionary pressure into the nervous system. The processes which are released in the brain by these systems are typically named feelings, emotions, motivations or needs. This comprises a multitude of types of excitations which have different purposes and which are experienced in a quite different manner. But there are also good reasons to see this as a unique system, which is in fact composed of multiple components, but which are also balanced to one another in a systematic way. However, since we try to understand the brain as complete system, what can only be achieved by some kind of abstraction and generalisation, it is suggested to emphasise the commonalities. That's why this aspect will be discussed hereinafter as "emotional-motivational system" (see next chapter).

In fact, are the signals and excitation patterns from these four systems bound into the other areas and processes of the nervous systems in a way which is quite similar. This occurs mainly via interconnections to cortical areas that are attributed to memory, behavioural control and cognitive control. These are mainly the memory regions entorhinal cortex and hippocampus as well as the regions which are called cingulate cortex and prefrontal cortex, and which are linked to decision making and control functions. Most of the pathways and loops of these systems are cross-linked via the structures of the basal ganglia. Thus, the basal ganglia is the crucial key component of the *emotional-motivational system* of the brain. Each type of feeling, emotion, motivation or need is conveyed via this intermediary device.

See also chapter "Emotion, motivation and memory" in Part 2.

The SPP model

This chapter develops a model of the brain as a system. This is done as combination of some abstract concepts, with the consequence that a lot of detailed findings is provisionally eliminated by generalization. Domain specific neuroscientific research leads regularly to a scattered concept of the human, composed of multifarious pieces of knowledge, but making holistic explanations of the brain as a system typically more complicated. The SPP model tries to counter this tendency by an attempt to compound the puzzle a little bit. The risk is that the degree of abstraction and synthetisation goes too far, so that the model is detached from evidence based scientific research. But there are strong indications that the developed concepts are not completely wrong, what is outlined in accompanying chapters and sections in Part 2, which are referred to in various passages in Part 1.

The findings and theses which are described in this chapter are primarily based on the following sources: Popper and Eccles 2006, Gazzanega and Mangun 2014, Singer 2004, Block 2009, Tononi 2012, Maslow 1987 and Kahneman 2012.

Referring to Singer 2004, it should be mentioned that this chapter is not consistent to his descriptions in all respects, but that some of his proposals have been adopted and others have been incorporated in a creative way. Referring to Kahneman 2012 it should be mentioned that only the basic idea of his book – about the "two systems" – has been incorporated in section "The two types of consciousness", but that this section is also supported by some of the other sources and notably by the other parts of the SPP model.

The emoti(onal-moti)vational system

A key point for the understanding of the brain is that it is an engine that follows purposes prior to the fact that it is an engine that processes information. And each kind of purpose is elicited by evolutionary pressure. That evolutionary pressure is conveyed into the brain processes is the only reason why the brain functions and is optimised in any way. Animals, humans and brains would not exist without purposes of this kind.

The evolutionary pressure takes mainly effect via the autonomic interface, involving homeostatic regulation, the sensory system, namely pain sensation, emotional processing, and the reward system. The corresponding excitation patterns are experienced as feelings, emotions, motivations or needs. This has been described above in section "Homeostasis, pain, emotions and rewards". An important finding was also that this comprises multiple components, which are indeed quite different, but which are bound into the other brain processes in a similar way, at which the basal ganglia takes effect as an important crossway in each case. A conclusion was that these systems have certain commonalities, so that it is adequate to generalize them under the concept "emotional-motivational system".

Another rationale for this generalized view is also that the brain is in the end a complete system, which must be able to balance all aspects of life to each other. This is only possible if there is a unique value system, which makes unified weighting of all types of emotions or motivations possible, independent from how different they might be. Systematic decision making and pattern selection is only possible if there is a common denominator available. This works only on the basis of a single currency, so a common emotional-motivational value system.

Consequently to this consideration, it might not be wrong to compress all types of feelings, emotions and motivational excitations to a common concept – we propose term "emotivation" for this aspect and will use it hereinafter consistently.

The consequence is also that there can be the talk of emotivational excitation patterns, emotivational values, and the emotional-motivational system can also be named emotivational system.

This concept, including the finding that there must be a common unified value system in the brain, is an important basis for the further concepts of the SPP model. Without this value system, the brain would not be able to balance all aspects of life with each other.

The control levels of the central nervous system

As most crucial topic, the SPP model tries to clarify how the human is controlled by his central nervous system. This occurs via several control levels which are functioning hand-in-hand and which represent different levels of complexity of the same coherent system.

Notes:
- Term control is always meant in the sense of regulation, feedback control, loop control or control circuits, but never in the sense of steering. Steering makes no sense, because the purpose of the nervous system is to regulate the interactions between human and environment for the benefit of the human, what means that natural laws must always be involved via feedback control.
- The descriptions of the first two levels can still be seen as a summary of indisputable basic knowledge while the

conceptual character of the SPP model starts with the descriptions of the higher levels (3), (4) and (5).

(1) Unconditioned reflexes

The most elementary control level is provided by unconditioned reflexes that are typically established by reflex arcs in the spinal cord. These are links between afferent nerve cords, receiving signals from specific receptors, and efferent nerve cords, stimulating specific muscle contractions. These reflex arcs can have excitatory or inhibitory character. They are important for all types of quick reactions that are always required in the same manner. The body with its limbs, receptors and muscles is determined by the DNA, and so are the unconditioned reflexes. Both body and appropriate unconditioned reflexes are no matter of plasticity in the individual, they are rather formed by the evolution via DNA variability in coherent manner.

An example is the interplay between antagonistic muscles. If an arm is bent by the biceps, it is always required that the triceps is released and vice versa. This is achieved by proprioception in the involved muscles and mutual inhibition of the appropriate motor nerve cords. The interaction of all muscles is coordinated by proprioception and well-adapted reflex arcs and control loops, at which only the most fundamental part is coded as unconditioned reflexes – for the other parts see the next control level.

Another example are protective reflexes, as the cough reflex, preventing impurity of bronchia and trachea, or the corneal reflex, protecting eyes against foreign particles.

There is a lot of protective and coordinative behaviour already coded at the lowest control level. This is a basic precondition for the ability to enfold more complex behaviours via higher control levels without larger risks and loss of efficiency.

Note: We are talking here about control level (1) of the *central nervous system*, but when dividing between central nervous system (CNS) and peripheral nervous system (PNS), control level (1) would refer to the latter; however, this is not very relevant, because there is no functional border between CNS and PNS; it is rather the case that the majority of the peripheral nerve cords is bound into the CNS.

Level attributes: Qualified perception: no. Emotional-motivational binding: no. Procedural memory: no. Declarative memory: no.

(2) Sensorimotor procedures

The next control level is provided by the sensorimotor control system. First of all, this comprises perception – by including all senses, afferent nerve cords and sensory cortices –, motor control – via the motor cortex, afferent nerve cords and innervations of muscles –, and the direct links between sensory and motor cortices. This system allows coordinated action on the basis of external and internal sensory information (e.g. sense of touch, proprioception), qualified perception (e.g. detection of objects by vision) and feedback control (e.g. detecting provoked modifications). Concept "procedures" means that this is about activity sequences which are composed of single activities, and which can be initiated, interrupted, varied and terminated at any time – why and how this occurs is explained by the next control level "voluntary movement" (see below). The procedures are encoded in the procedural memory in terms of well-trained behaviours, habits and conditioned reflexes which can be exploited by the higher control levels for any purpose. (Note: levels (1) and (2) are seen as not involving purposeful components directly on their own; the purposeful components of habits and conditioned reflexes are incorporated via level (3).)

Level (2) comprises also a complex neural infrastructure which is required to turn intended actions into smooth well-regulated movements and which supports body control in general. Characteristics of this type can be attributed to subcortical structures and namely to the cerebellum. This includes features like tonicity, posture and balance.

Qualified perception, as e.g. detection of objects by cutaneous perception and vision or differentiation of tone pitches and sequences, is included as crucial ingredient at all control levels, starting from (2).

Level attributes: Qualified perception: yes. Emotional-motivational binding: no. Procedural memory: yes. Declarative memory: no.

(3) Voluntary movement (activity control)

The voluntary movement level adds decision making, pattern selection, impulse control, vigour control, modulation and multiplexing to the level-two procedures and level-one reflexes. And this is the first level which binds excitation patterns into the system that represent emotions, motivations, needs or purposes. The most crucial cerebral areas representing level-three characteristics are the basal ganglia (BG) and the anterior cingulate cortex (ACC). The mechanisms of control level (3) are functioning in close cooperation with control levels (4) and (5), at which both areas play also an important role.

Decision making occurs at all levels starting from (3). But level (3) is associated with the primary type that might be called affective or impulsive decision making. This means that a specific stimulus or a combination of stimuli attracts attention and triggers connected sensorimotor procedures immediately. This is connected with concepts like stimulus-response association, conditioned reflex and habit. These types of direct reactions to stimuli or of habitual behaviours can be learned, adapted and also dissolved again at any time.

The decision making process has two aspects:

- paying attention on an internal or external stimulus and adopting it as a need (goal decision making, primarily done via the basal ganglia),
- and selecting a sensorimotor pattern to be excited and executed as starting point for the reaction (sensorimotor pattern selection or motor decision making, respectively, mainly processed in the anterior cingulate cortex).

The first reflex may then trigger further sensorimotor sequences, so that longer habitual behaviours can be triggered by this mechanism.

Following the characterisation until her, this seems like talking about impulsive or compulsive behaviours, while it should actually be attempted to explain the opposite, according to term "voluntary" in the level name. So, how can this type of mechanism turn into "voluntary movement"? Via the involvement of four specific basal ganglia circuits that are crucial for the ability to stop impulsive reactions and compulsive sensorimotor sequences, to start the search for alternatives and to influence and modulate all kinds of behaviours in a more regardful, controlled and even voluntary way.

That impulsive/affective/compulsive behaviour may turn into voluntary/regardful/controlled behaviour is the result of the interplay between levels (3), (4) and (5), while level (3) alone would not have this potential; impulse control (level three) would not make sense without the search for alternatives (level four/five). But the fact that higher control potential is basically incorporated makes the characterisation "voluntary" appropriate for level three. How this potential is exploited is outlined in the descriptions about levels (4) and (5) below.

However, the crucial basis for all these performance features is provided by four basal ganglia circuits, as mentioned above. "The basal ganglia (BG) is composed of several heavily interconnected nuclei at the base of the cerebrum. A series of anatomically distinct parallel circuits through the BG receive afferent projections from and project back to cortical regions that mediate skeletomotor, oculomotor, frontal associative, and limbic functions." (Turner and Pasquereau 2014, 435).

These four basal ganglia circuits are interconnected as follows (ibid. 436):

- the skeletomotor circuit with the motor cortex, controlling sensorimotor functions for the body,
- the oculomotor circuit with the frontal eye cortex, controlling eye movements,
- the associative cortex with the dorsolateral prefrontal cortex, representing cognitive functions and
- the limbic circuit with the anterior cingulate cortex, representing emotions.

The specific control functions that are attributed to these four circuits, can perform independently in parallel. Besides this, these circuits are also integrated by some kinds of "projection", "communication", "convergence" and "funneling". Pathways are mentioned, for example, which support projections from limbic regions to nonlimbic frontal cortical areas or to motor cortices.

The result is that oculomotor control and associative reflection takes place in parallel to and widely independent from skeletomotor control. One can move into one direction while his eyes and thoughts wander into two other directions. Another result is also that the influence of emotional-motivational excitation (limbic) can always be regulated in relation to action (motor) and thinking (associative).

The most interesting and also best analysed circuit is the skeletomotor circuit. It is connected among other things with the characteristics impulse control/intervention, vigour control, modulation and reinforcement-driven skill learning. Impulse control means that impulsive reactions, once selected via the affective decision making mechanism, can be intervened again very quickly via a special "hyperdirect pathway" so that there is no sign outside the individual which could be an indication for the actual impulse. Vigour control means that amplitude, speed and force of movements can be controlled through the impact of emotional-motivational excitations. Modulation is continuous fine-grained vigour control for each small part of sensorimotor sequences. Skill-learning is only possible if well-trained behaviours are partially disrupted (via control level three) and varied (via control levels four and five), so that new opportunities and sensorimotor patterns can be developed. The intervention and vigour control mechanisms in the basal ganglia are crucial for all these capabilities of the motor control system.

These great abilities are at the same time accompanied with large potentials for malfunctions. Parkinson's disease, for example, is attributed to troubles with the skeletomotor circuit (namely with the dopamine system, which is involved here), and smooth habitual motor activity is only possible without the excessive interventions in the basal ganglia, that are a typical complication of Parkinson's disease. See also Turner and Pasquereau 2014, 440ff.

How all these mechanisms exactly work, is still a matter of research. But it is known that they work and where they work. However, the basal ganglia comprises a great multiplexer component with parallel processing and intermediation abilities between motor control, associative-cognitive functions and emotional-motivational influences on the one hand, and between control levels (3), (4) and (5) of the nervous system on the other hand.

On a summary, voluntary movement means making decisions about the right reactions and habitual behaviours due to perceptual excitation patterns (external stimuli) and emotional-motivational excitation patterns (internal stimuli), and to do this always with the ability to interrupt or prevent these reactions if recognised as inappropriate for any reason. Seen on a more detailed perspective, this comprises also control and fine-grained modulation of movement vigour in terms of amplitude, speed and force.

Based on the assumption that there are other (more qualified) decision making mechanisms provided by the higher control levels (4) and (5), it is also the task of the voluntary movement level (3) to adopt these decisions and proceed with the selected sensorimotor patterns, just as if these behaviours were directly initiated by the level-(3) mechanisms, and always with the potential to interrupt it and return the control again to the higher levels.

For more details about the basal ganglia circuits see also chapter "Basal ganglia (BG) and frontal cortex" in Part 2 and Turner and Pasquereau 2014.

Level attributes: Qualified perception: yes. Emotional-motivational binding: yes. Procedural memory: yes. Declarative memory: no.

(4) Situational preparation

It would be ideal if reflexes and habits worked always in a suitable way to the benefit of the individual. Then, control level (3) would be enough. But there are many reasons why affective and habitual movements may become inappropriate. Activity is stopped then, and the search for alternative reactions or behaviours, so sensorimotor patterns, is started. This is the task of control level (4). It has major implications.

Firstly it must be clear that control levels (3) and (4) are not working alternately. It would be odd-looking if beings act and stop acting (while thinking) in turn. It is rather typical that this occurs in parallel. Each type of activity sequence can be seen as a continuous series of junctions between possible alternatives to proceed. So it is only logical that future decisions are evaluated by one type of neural processes while activity control is ongoing by another type of neural processes. The precondition that this can occur in parallel is a neural system that can manage both types of processes in segregated systems. As outlined above, the brain is equipped in this way, at which the four basal ganglia circuits, as described at level (3), play a crucial role. Activity control is executed by the skeletomotor circuit while preparation of the next decision(s) occurs at the same time by the associative-cognitive circuit.

The level-four decision making system is described in the literature as "value-guided decision making" (see in Rushworth et al. 2014, references therein and section "Decision making in the frontal cortex" in chapter "Basal ganglia (BG) and frontal cortex" in Part 2). It is attributed to specific structures in the prefrontal cortex (particularly ventromedial prefrontal cortex and the adjacent medial orbitofrontal cortex) and to the anterior cingulate cortex. The frontal cortex areas are known for making decisions between different possible reward goals by evaluating reward magnitude (size) and reward probability. This occurs, for example, as binary decision between two different reward goals that are present, but also as search/engage decision between a present reward goal and better opportunities in the future. Like all other types of decisions (and pattern transformations), these decisions are processed by synaptic multilayer networks that are well-trained for this purpose. It can be assumed that a decision occurs regularly as maximum function of a neural multilayer map in a grey-matter module, so that excitation is finally concentrated to a specific bundle of output neurons. This information is then transferred via white-matter axons to further modules and brain areas, where it takes effect for successive stages of the neural processes.

Once a decision for a reward goal is made, the next step is to select an appropriate activity pattern for achieve the reward goal. This task is known as typically done in the anterior cingulate cortex. But this is now again a common ground between levels (3) and (4), i.e. this part of the decision-making cascade occurs also for affective level-three decisions in the anterior cingulate cortex, so that only the reward-goal decision part is the real unique feature of the value-guided level-four decision making processes. The result is that control level (4) represents in particular weighing between opportunities, like different reward goals and needs or the question whether to proceed with activity towards a specific goal (engage) or classify this as inappropriate and intervene (search).

Regarding decision making processes via cascaded and diversely connected neural modules and multilayer maps, Singer 2004 points to the fact, that multiple competitive processes are typically involved, and to the phenomenon, that slight variability in signal processing can result in different decisions (see in Singer 2004 under headline "Different forms of knowledge"/"Verschiedene Formen des Wissens").

These decisions are processed in close cooperation between the frontal cortex, the anterior cingulate cortex, the basal ganglia and the emotional-motivational system. The frontal cortex selects the reward goals and needs which will have priority, decides what the next goal is and whether activity has to proceed or must be intervened in favour of a more complicated search for alternatives. The decision result takes effect via the multiplexer in the basal ganglia (e.g. via the skeletomotor circuit). The anterior cingulate cortex selects and manages the activity sequences consequently to the goal which has been prioritised in the basal ganglia (impulsive/affective) or in the frontal cortex (value guided). The emotional-motivational system supplies the values which are always the driving factors for these processes.

A further task of the frontal cortex is also to manage the handover between affective/habitual behaviours and appropriate value-guided forecasts that might be performed in parallel for their support, and, if necessary, the interruption of activity in terms of impulse control for deeper preparation; the anterior cingulate cortex and the basal ganglia circuits are involved in this kind of operations. Another task of the emotional-motivational system is also to bind vigour control and modulation into activity sequences, what takes effect via the basal ganglia circuits.

At level (4), the focus for all these processes is always the current situation of the individual. The general aim is to act in the current environment for the own benefit. Forecasts and value-guided considerations about alternative goals and activity patterns concentrate on the current spatiotemporal surroundings and opportunities. It is clear that humans go mostly far beyond this limitation and reflect often experiences and opportunities that are far in space or in time, but this is a contribution of level (5) and level (4) is only involved in this via close cooperation with the level-five processes.

Declarative memory does not make sense until the affective/voluntary level (3). Conditioned reflexes and habits follow a determined schema of direct interactions with the environment, so that there is no room for distant positions in space and time or for alternatives. But this changes with level (4). Preparation and value-guided decision making means involving similar experiences in the past and at different locations, comparing them and developing ideas (alternative candidate patterns) of what may happen due to which potential decision. This requires some basic kinds of memory and cross-modal associations and the ability to involve this into the evaluation processes. That's why the development of declarative memory may have started with the first types of level-four processes at any stage in the phylogenetic history. The powerful declarative memory system as it is developed in the human brain today goes far beyond these requirements - this is clearly an outcome of control level (5) (see below). But because situational preparation processes at level (4) execute typically in close cooperation with level-five processes, it stands to reason that the full capabilities of the memory system can also already be involved in level-four control processes.

Level attributes: Qualified perception: yes. Emotional-motivational binding: yes. Procedural memory: yes. Declarative memory: yes.

(5) Creative preparation

Control level (5) bases still on value-guided decision making and on the same circuits and loops in the prefrontal cortex, anterior cingulate cortex and basal ganglia as exploited on level (4), but it goes beyond the spatiotemporal scope of current activity. It involves interpretation of past experiences, forecast to future activity and implications over a larger spatial range. It corresponds with means that can be used for planning and modelling (e.g. paper and pen) and with communication and interaction in societal contexts. It leads not always to a result for initiate the next activity, but rather to kinds of long-term preparation for hypothetical activities that might become due in the future. Conclusions and decisions for current activities are naturally also always possible.

Association and cognition without spatial and temporal limit is closely connected to the development and exploitation of declarative memory. Target oriented movement towards far locations require complex spatial maps in the hippocampus. The ambition to recognise structural object relationships and interrelations in complex processes require cross-modal associations with no limit; difficult challenges can only be tackled if using all senses in cooperation, or even through the development of additional senses by technology (e.g. infrared sensor, radar, microscope, magnetic resonance imaging). Planning requires the perception of time as well as linking memorised experiences and new ideas with time scales, what is facilitated via the entorhinal cortex (see also section "The types of memory" in chapter "Emotion, motivation and memory" in Part 2).

By this way, creative preparation opens the path to internalise a world model, organise collaborative activity and to take control over processes and developments in the external world.

Language plays a particular important role in the level-five control processes. It is clear that language is developed for communication purposes. It can be used to coordinate activity and adjust world models between individuals; with libraries and modern communication technologies, this works not only in direct team contexts, but also in societal contexts over large spatial and temporal distances – virtually from the first book in history over the other side of the globe and the outer space to the here and now.

But language has, besides communication, another important aspect. It is also the most significant means of modelling of the world, which takes also effect when not acting directly in societal contexts. We are social beings to such an extent, that we, even if we are alone, process associative and cognitive patterns in a way as if we communicate with an imaginary counterpart or with ourselves. This is not only true for spoken or written language, but also for other means of communication like used in art and engineering (e.g. pictures, mathematical models and so on).

Returning to the neural processes in frontal cortex and basal ganglia, it is likely that linguistic capabilities are mainly linked to the associative/cognitive circuit of the *multiplexer* in the basal ganglia – see also Chan, Ryan and Bever 2013 and section "Basal ganglia (BG) and language" in chapter "Basal ganglia (BG) and frontal cortex" in Part 2. This means that language might be an inherent ingredient of associative and cognitive processes rather than of skeletomotor activity control processes. Another consequence is that the development of cognitive capabilities is, in return, closely bound to the development of language (and of other forms of communication). There are few results of associative thought that are not devoted to communication at the same time in any form.

In relation to *creative preparation*, it can be stated that written and spoken language is implicitly a kind of associative/cognitive process, and that *creative preparation*, if ongoing internally, might partially be a kind of linguistic simulation, i.e. it is silent speech. The vigour control mechanism in the basal ganglia plays the crucial role for switching between quiet simulation and loud articulation. An important result of this view is also that spoken language and thinking can only be processed simultaneously, while skeletomotor and oculomotor activity control occurs in parallel to thinking. This makes speech (e.g. at a presentation) the most difficult control task on level (5); short pauses must be exploited as good as possible to remember the context and to plan the next sentences, and a good speech is usually based on thorough preparation.

What is traditionally specified as short term memory is nothing else than the constantly ongoing control processes in basal ganglia, frontal cortex, anterior cingulate cortex and in the memory system. The memory system is insofar involved as value-guided decision making means evaluation of experiences and associations, what means memory recall. Recall or reinstatement of declarative memory content does not occur without evaluation, and evaluation means nothing else than assessment of declarative memory content in relation to a goal and, in case of situational preparation, also in relation to perceptions from the external environment.

The control processes can short term switch between various goal-pattern combinations within the time frame of less than one minute, what is traditionally attributed to the concept short term memory. But in reality it is continuous intermediation between emotional-motivational patterns and goals on the one hand and memories, sensorimotor patterns and perceptions on the other hand.

A peculiarity of level-five control is also that hypothetical evaluation processes, which do actually not lead to visible results by visible activity (so to physical results in the environment), can turn into kinds of situational preparation and of activity (levels four and three) by developing and exploiting means for the expression of intermediate data. Most of human activity fits to this category. Planning, engineering, bureaucracy, science, journalism etc. produce nearly nothing than intermediate results that can be communicated. Human activities which lead directly to final results (e.g. products) are nowadays more an exception while external representations of associations and ideas that develop in the cortex are the rule. This kind of development might not occur without reason, since level-five control is a social achievement, which bases significantly on communication, and there might be, respectively, a general inherent pursuit to exploit lower control levels prior to higher control levels or to reduce complex problems – like the long term aspiration of a goal –, to easy solutions – like immediate production of intermediate data (see also section "Efficiency through delegation and structuring"). Language (and each kind of communication) is, in the light of this view, again an exceptionally interesting feature – it is a great bridge between both level-five associations and level-four-to-three activities, typically maintaining representations of the former in the external world (texts, figures, videos etc.).

Note: Control levels (1)-(4) may correlate with *System 1* from Kahneman 2012, and control levels (4)-(5) with *System 2* from the same book. More details see in Part 2, chapter "Psychology", section "Kahneman, Daniel: Thinking, Fast and Slow", and in Kahneman 2012, "PART 1. TWO SYSTEMS", 17ff.

Level attributes: Qualified perception: yes. Emotional-motivational binding: yes. Procedural memory: yes. Declarative memory: yes.

The attention assessment controller (AAC)

The control levels (3) voluntary movement, (4) situational preparation and (5) creative preparation are working in a highly integrated way. The brain areas and modules which facilitate these types of control, so basal ganglia, frontal cortex, anterior cingulate cortex and the emotional-motivational system are closely connected to each other and to the cerebral cortex by strong pathways and loops. Some processes, like skeletomotor control, oculomotor control and associative processes occur in parallel, but the overall working principle is coordinated interaction and steady integration and handover between the three levels.

So it is adequate to combine these levels under a joint concept – let us call it "attention assessment controller" (AAC).

What does this stand for?

Attention means decision making and pattern selection in two regards:

- selection of a stimulus or a reward or a need, respectively, which turns into the prevailing goal,
- and, subsequently, the constant sequential selection of associative/cognitive or sensorimotor patterns, which are excited for approach the goal step by step.

This occurs in three modes, according to the control levels (3), (4) and (5): activity control, situational preparation and creative preparation.

In this context, attention is meant in the neurophysiological sense, so in terms of *active* pattern selection and decision making processes, as outlined above, rather than in the sense of (passive) psychological attention towards a perceived stimulus (like a whistle or a speaker's voice). However, the latter can also be seen as one of the observable effects of the former.

Note: The book is here that attention towards a perceived stimulus is only possible if any kind of internal state is active or potentially active that classifies the stimulus as relevant. This can be any emotional-motivational pattern, so a need that is either already excited or that is in standby. Perceived signals that do not meet an emotional-motivational pattern or a need are hindered in becoming stimuli and thus are regularly ignored.

Assessment means that the excitation patterns that are circulating in the system, so in the areas that are attributed to the control levels (3), (4) and (5), are constantly evaluated and assessed, and that they are only potentially persisted on the basis of the resulting values. This evaluation and assessment process is the primary mechanism for establish patterns in the memory system, namely in the declarative part and indirectly also in the procedural part.

It can be supposed that this occurs in two respects:

- As association between informational patterns on the one hand – e.g. perception patterns, microrepresentations of sensorimotor patterns, spatial patterns, patterns representing time – and emotional-motivational patterns on the other hand – e.g. patterns representing autonomic states, pain, emotions or rewards. This occurs in the association cortex and primarily in the entorhinal cortex, which is seen as hub to all other areas of the memory system and as multimodal association area, supporting unspecific associations between various modules and association areas with more domain specific determination. Moreover, the entorhinal cortex is an area of the memory system where emotional-motivational signals are directly in-bound by strong pathways from lower regions, namely via the basal ganglia. This area is also called limbic association cortex. See also chapter "Emotion, motivation and memory" in Part 2.
- As excitatory and inhibitory effects at synapses in all brain areas, at which excitations represent positive values and inhibitions represent negative values. The associativity and synaptic plasticity in the whole brain is constantly developed and optimised for the service of the human species. This occurs at all synapses of the brain in a more or less indirect way via the primary mechanism from above.

Summarising, one can state that evaluation and assessment is the (only) mechanism for establish memory content in the brain and that there is no memory without emotional-motivational value. The finding, that the brain is constantly optimised by the assessment process, might not only be true for the networks of synapses (the "software"), but eventually also for the long-term development of the whole brain structure and of the body (the "hardware"), since what is assessed is the performance of the complete organism.

Note: The concept *attention assessment controller* might not be very far from what is traditionally called limbic system, but the latter has got various meanings over the time and the former is introduced and preferably used here for the ability to make better clear what is meant.

The concept *attention assessment controller (AAC)* will now be outlined in more detail with the help of figure 1, which shows the crucial elements of this concept. This representation bases on scientific findings about structures, pathways, circuits and control mechanisms in the brain, and it should be considered that it is an attempt to summarize the various findings within a simplified set of abstract principles. But simplification must not be wrong in general, since nature is often based on simple principles which seem only complicated as long as they are not well-understood. The summarising view, which is provided by figure 1, and the explanations below are an effort to not lose the sight of the essential aspects, despite the extraordinary complexity of the neurological matter.

A condensed abstract perception, which combines the most important aspects in hypothetical manner and which ignores a lot of secondary aspects, can be seen as useless by domain experts, who focus on the highest level of detail. But it might be a necessary supplement to domain specific knowledge, to also ask for holistic, systemic or interdisciplinary views.

The pillars of the AAC concept are the interplays between "assessment" and "attention control" as well as between "activity mode" and "preparation mode". The descriptions of the brain structures and processes from above are now summarised again in accordance to these concepts.

Assessment

Assessment means continuous evaluation and assessment of the results that are achieved by the control processes of the brain. The general aim and purpose of human activity is always survival and thriving as social being under the conditions of evolutionary competition. This purpose is bound into the brain processes via the emotional-motivational system, comprising patterns from autonomic interface (including homeostasis), pain sensation, emotional processing and reward system (see box **"Emv"**/"Emotivations"). The consequence is that storage of excitation patterns in the memory system takes always effect on the basis of emotivational values that are emitted by the emotional-motivational system. These values are an inherent ingredient of each type of memory content, at which this occurs as emotivational association in the association cortex and as excitatory and inhibitory effect at all synapses in the brain. The emotivational value, which applies to an informational pattern – representing perception, sensorimotor procedure, space or time etc. – can be gradual positive or gradual negative. It can be assumed that there is no memory content, no synapse and generally no part of the body which is not assessed and endued by a value by this way.

Assessment is represented by the two assessment symbols in figure 1. Perception and sensorimotor activity is always assessed directly (see the **"Asm"** symbol at **"SC"**/"Sensory cortex" and **"MC"**/"Motor cortex"). This basic stuff happens always, also if the individual follows level-three reflexes and habits without thinking twice, i.e. without impulse con-

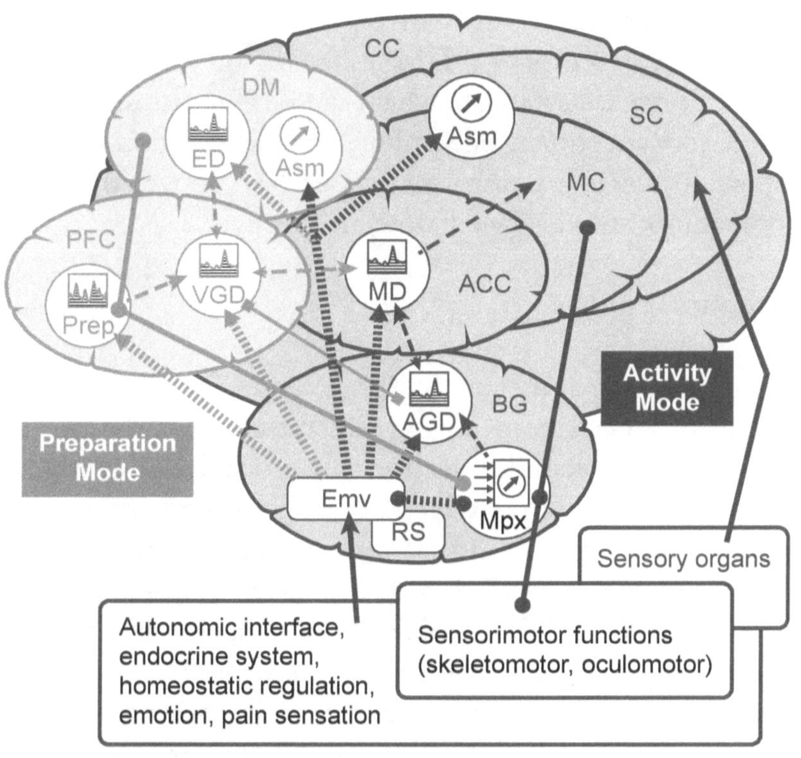

Functions

AGD   Affective goal decision
Asm   Assessment
ED   Encoding decision
Emv   Emotivations
MD   Motor decision
Mpx   Multiplexer
Prep   Preparation
RS   Reward system
VGD   Value-guided goal decision

Areas

ACC   Anterior cingulate cortex
BG   Basal ganglia
CC   Cerebral cortex
DM   Declarative memory
    (association cortex, tem-
    poral lobe, hippocampus)
MC   Motor cortex
PFC   Prefrontal cortex
SC   Sensory cortex

Figure 1 – Attention Assessment Controller (AAC)

trol or appreciable level-four or level–five incidents ("level" means control level – see above; level three refers also to the "activity mode" and level four and five refer to the "preparation mode" – see below).

The other **"Asm"** symbol represents assessment in the "Declarative memory" (**"DM"**). This is the mechanism of how experiences establish, which can be consciously remembered and which can be exploited via association processes, level-four/five control processes (see above) or by the "preparation mode" (see below), respectively.

Attention control

Attention control means, first, selection of a stimulus or a reward or a need, respectively, which turns into the prevailing goal, and, second, selection of associative/cognitive or sensorimotor patterns, which are excited for approach the goal step by step. The first is represented in figure 1 by the two goal decision symbols (see **"VGD"**/"Value-guided goal decision" and **"AGD"**/"Affective goal decision"), the second by the **"MD"** symbol ("Motor decision").

The trigger for a decision can elicit from inside or from outside of the organism. Internal triggers are, of course, signals from the emotional-motivational system. Needs require attention and reactions, and this cascade is processed via goal decision (**"AGD"**/affective, **"VGD"**/value-guided), motor decision (**"MD"**) and resulting behaviours (see the three symbols in figure 1). During the released activity, the same functional chain works also as control loop for continuous corrective control.

External triggers can occur via any perception pattern. But perception takes place through a complex cascade of filters (mainly in the sensory areas), whereat the last and most important filter is the emotivational filter – only if a perception pattern meets any kind of interest, in terms of an emotivational pattern that is in standby, it can happen that an external stimulus turns into a goal decision.

Affective goal decisions (**"AGD"**) are the standard, which has been developed long ago in the phylogenetic history. This refers to control level three (see above) and to the activity mode (see below). Value-guided decisions ("**VGD**") occur when impulsive or habitual behaviours are detected as not suitable, what happens the more often the higher developed the individual and the more exploratory the current course of action is. This refers to level-four/five control (see above) and to the preparation mode (see below).

The crucial point of attention control is the evaluation process. The basic principle of each kind of evaluation and decision is comparison of concurrent value-added neural patterns, which may come from similar or rather different sources. The first stage – goal decision making (see the **GD** symbols in figure 1) – compares concurrent emotivational patterns which are sent from the autonomic interface, pain sensation, emotional processing or from the reward system. These patterns are comparable, and one of it gets temporarily the priority, what is a more or less stable result, but which can change at any time and which can switch between multiple goals with high frequency.

The second stage – motor decision making (see the "**MD**" symbol in figure 1) – compares the selected goal, providing a demanded value, with value-added solution patterns that are available in the memory system and which have got their values by the assessment process during storing. This can be multiple concurrent sensorimotor patterns, or it can be one or no pattern (in consequence of informational pre-filtering). The pattern is selected in the end, which meets the demanded value best.

The second stage – motor decision – can also backlash towards the primary stage – goal decision. This may happen directly, e.g. if no sufficient solution is found, or via a detour through the triggered sensorimotor procedure and another intervention in the BG multiplexer ("**Mpx**"), forcing further decision making procedures.

However, this works similar to other control mechanisms. The evaluation process or the complete functioning of the attention assessment controller can be seen as just another variant of a control loop with nominal-actual value comparison. At the decision making cascade, the nominal value is emitted by emotivational patterns representing needs, and the actual values are the selected solution patterns. The system is balanced if activity is ongoing on the basis of stable goal and habitual behavioural sequence. So it is a kind of activity regulation system. Intervention is necessary if activity is inappropriate.

For the whole attention assessment controller (AAC), the set value might be the total absence of deficiency needs, and the actual value is provided by any kind of deviation from this general aim. Or, along with rewards and artificial/virtual needs, the set value is aesthetic experience and the actual value is its absence, and balance is provided if no deficiency need is active and, possibly, if aesthetic experience is ongoing as demanded (the role of artificial/virtual needs and aesthetics is explained below, starting with section "Higher needs"). Intervention is necessary if needs occur, so nearly always. The result is that the AAC makes the brain a great self-regulation system, which regulates human affairs in the process of evolutionary competition.

All the goal decisions and sensorimotor decisions occur as *suitability probability* evaluation processes in synaptic multilayer networks or decision making maps, respectively. What this means is described in section "Universal suitability probability evaluation" below. A short explanation of *multilayer networks* and *decision making maps* has already been provided in chapter "Neuro basics" above.

It is important to note (again), that term attention is meant in the sense of "neural attention". This has little to do with the well-established concept of attention towards an external stimulus, as, e.g., a whistle, a speaker's voice or a concert. Term attention is here used in the sense of attention towards neural patterns, or filtering and selection of neural patterns and turning excitation processes towards these patterns. The conventional well-established attention concept is only one of the observable phenomena, which may occur due to neural attention.

Activity Mode

The activity mode is linked to level-three control, the skeletomotor and oculomotor circuits through the basal ganglia (see symbol "**Mpx**"/"Multiplexer" in figure 1 with the connection between "**MC**"/"Motor cortex" and "**Sensorimotor functions**"), affective goal decisions ("**AGD**") and motor decisions ("**MD**"). How these loops for direct activity control work is determined by informational and emotivational filtering of memorised neural patterns.

Informational filtering means alignment of neural excitations to the informational state of the individual in the environment. This refers to perception and the influence of perception results on decision making and pattern selection processes (see "**Sensory organs**" and "**SC**"/"Sensory cortex"). This includes external perception, as vision, sound, touch, taste etc., and internal perception, as, e.g., proprioception.

Emotivational filtering means alignment of neural excitations to internal need states. This is managed via the influence of the emotional-motivational system in the basal ganglia circuits (see the connection between "**Emv**"/"Emotivations" and "**Mpx**"/"Multiplexer") and via "**AGD**"/"Affective goal decision" and "**MD**"/"Motor decision", which are controlled by emotivational values as described above under "Attention control".

These **"Activity Mode"** control loops build the fundament on which higher-order control mechanisms, as level-four/five control and **"Preparation Mode"** can evolve.

Preparation Mode

The preparation mode is linked to level-four/five control, the associative circuit through the basal ganglia (see symbol **"Mpx"**/"Multiplexer" in figure 1) and associative processes that are mainly controlled by areas in the prefrontal cortex and that are able to exploit and influence declarative memory contents (see symbol "Prep"/**"Preparation"**). This is a great play with all the options that are available in the neural pattern library, primarily in the declarative part, what includes that different contents can be combined to one another in any form, which is not predetermined.

This is an excitation control process, which is continuously ongoing in the brain independently from and in parallel to sensorimotor and oculomotor control processes, what is mainly a merit of the four basal ganglia circuits that are represented by symbol **"Mpx"**/"Multiplexer" in figure 1 (see also the descriptions under "(3) Voluntary movement (activity control)" in section "The control levels of the central nervous system").

These associative processes can be seen as a kind of information processing, but which is certainly lead by purposes and values. They follow the principles of *attention* and *assessment*, just as this is true for activity control processes. This includes selection of a (hypothetical) goal, evaluation of associative patterns that might be consistent to the goal and selection, assessment and storage of resulting patterns. This is a constant evaluation process which reorders and recombines the neural patterns in the declarative memory continuously and which produces always completely new representations of pattern combinations in the associative areas. This is the type of process, which can be experienced as conscious thinking under certain circumstances (more details see below in sections "The two types of consciousness" and "Conscious experiences").

The potential to interact with the control of real activities is always included. The associative preparation processes are sometimes completely detached from the sensorimotor control processes, so in the *creative preparation mode* (control level five), but they can also be closely linked to the current activity context in the sense of the *situational preparation mode* (control level four). In the latter case, goal decisions are not always only hypothetical, only influencing internal associative processes, but they can rather also become real in the sense of the decision making apparatus for sensorimotor behaviours. This aspect is represented by symbol "**VGD**"/ "Value-guided goal decision" in figure 1. This type of decision making occurs always in close cooperation with the functions "**AGD**"/"Affective goal decision" and "**MD**"/ "Motor decision", what is supported by strong direct pathways between the appropriate areas in the brain, and what is represented by related connectors and arrows in the figure. The primary mechanism for this interplay is impulse control due to inappropriate habitual behaviours, what is processed in the basal ganglia circuits by (quick) inhibitive intervention of sensorimotor activity and as handover to deeper value-guided decision making processes, before more suitable behaviour can be started again.

It is an interesting finding that linguistic capabilities and processes, although connected with motor control of vocal cords and eyes among other things, can be mainly attributed to the associative circuit. There is a large distributed network of language in the brain, including a ventral semantic stream, a dorsal phonological stream, a speech perception pathway, and an articulatory loop. Due to current scientific findings, this network corresponds more with the associative basal ganglia circuit rather than with one of the motor control circuits (see also section "Basal ganglia (BG) and language" in chapter "Basal ganglia (BG) and frontal cortex" in Part 2). This suggests that language is close to association and preparation and, vice versa, that associative and preparative processes are significantly linked with linguistic processes. This is no coincidence, because it is clear that thinking, association, planning (preparation) is mainly cultivated in social contexts, and thus always based on communication. Human's understanding of the world develops in multimodal associations, whereat naming and linguistic communication plays always a crucial role. This means also that language is a great phenomenon, which provides external representations of parts of the processes that are internally ongoing in the associative areas with the behalf of the associative basal ganglia circuit.

Asking for the relation and priority between preparation mode and activity mode, it can be asserted that the preparation mode refers to reason and the activity mode to action, impulsive/habitual behaviours and instincts. A conclusion could be that well-controlled behaviour, which is always based on the detour over the value-guided decision making system, should, along with the development of culture, become more and more the rule and that impulsive behaviour should become more and more the exception. This perception is totally wrong.

The truth lies rather in a close and well-organised cooperation between both systems. Decisive association processes and considerate behaviours can only enfold when the processes of this type are not always exploited for demands that can be solved by routine behaviours. That's why the strategy must always be to develop bunches of more suitable behavioural sensorimotor patterns for tasks that may occur repeatedly in similar manner. Only if this pursuit is cultivated in distinct manner, it is possible to get enough leeway for the ability to manage real difficult tasks via the preparation mode in a good manner. A rich library of sensorimotor patterns is the crucial basis for high effectiveness of the whole system, at which few small value-guided impulses are enough for steer action always into a successful direction.

The preparation mode, on his part, is typically aligned with the general pursuit to feed the library of routine behaviours with as many suitable patterns as possible, or at least as necessary. This might be an inherent instinct of the associative control machinery, which occurs for its own relief and its ability to remain efficient. This is why our life is often more determined by learning and training than by effective action and why brain growth occurs to a decisive extent for complex sensorimotor processing, besides declarative memory and higher-order control mechanisms (see also the next section below).

Note: The *activity mode* and the unconscious part of *situational preparation* may correlate with *System 1* from Kahneman 2012, and the conscious part of *situational preparation* as well as *creative preparation* in general may correlate with *System 2* from the same book. More details see in Part 2, chapter "Psychology", section "Kahneman, Daniel: Thinking, Fast and Slow" and in Kahneman 2012, "PART 1. TWO SYSTEMS", 17ff.

That's it about the attention assessment controller concept. This is a model of the major control mechanisms in the brain, which might be little bit too brave, which is not free from synthetic ingredients and which will definitely need further critical review. But, on the other side, it would be a failure to not try such kind of complement to the usual neurological research strategies, which occur in highly specialised manner and which lead to a certain degree of blindness for greater interdependencies and for the holistic and systemic aspects of the brain.

Nature is often based on very simple, but efficient concepts. It seems often confusing to us because we are not able to understand the brilliancy of these concepts. Regarding the brain, there are two opportunities: deploring that it is too complicated to understand exactly how it works as complete system, or trying interim concepts, which may not be finally correct, but which reflect the suspicion that there must be ingenious intermodal concepts behind these endless bunches of nerve cords and synapses. The SPP model is an attempt to cultivate the latter idea.

The biggest open question:
Regarding the capability of the basal ganglia to manage skeletomotor control, oculomotor control, associative processes and limbic processes widely independently in parallel by segregated circuits, it is a big remaining question, of how the involved cortical areas are able to realise these divided processes on the first hand, to interact partially on the second hand and to do this without indulging in total confusion and chaotic flow of excitation on the third hand. See also "Question 1" in chapter "The biggest open questions" at the end of Part 2.

Efficiency through delegation and structuring

The processes of interaction and handover between control levels (1) to (5) include a constant natural pursuit for efficiency. The higher levels, particularly level (5), are mainly associated with all kinds of trouble or complication, while the lower levels (1) to (3) are rather associated with smooth behaviours, pleasant habits and well-trained skills that can be used without greater amounts of conscious attention. Of course, higher-level ambitions lead also regularly to satisfying achievements and aesthetic elations in the end. But this is a complicated detour that is only taken when unavoidable. The natural pursuit of the brain is rather being always competent and successful without greater mental efforts in terms of conscious reflection.

In fact, it is never possible to manage life mainly by level-five efforts. This approach would only lead to disasters, because explorative behaviours consume always too much time and too much other resources. The whole system can only act in sufficient manner if level (4) and particularly level (5) are reserved for exceptional cases and for extraordinary issues, while the majority of the daily requirements is successfully managed by well-trained lower-level skills. That's why the system acts in a way, so that each activity, habit and attitude is constantly optimised by feedback control. The effects and results, that are achieved, are always incorporated into the memorising processes as values, leading to the constant adaptation of sensorimotor-emotivational patterns and associative-emotivational patterns (so of the solution patterns), with the result that improvement occurs in the neural pattern library always as implicit ingredient of activity and preparation processes. Acting and thinking means implicitly always reinforcement-driven learning and refinement of all parts of the solution pattern library.

This path of constant optimisation occurs not only between the levels, but particularly also inside of level (5), including world models, communication processes and intellectual ambitions. To solve long-term problems in an intellectual way means always to find simple solutions for difficult problems, opening the room for manoeuvre for facing further issues with higher complexity.

Only if the brain structures and contents are optimised in this way and tasks are regularly delegated from higher to lower control levels, or from complex to simple solutions (inside of level five), the brain will have enough remaining higher-level resources when extraordinary issues have to be managed additionally. It is a natural quality of the brain, that it fits this demand always in widely sufficient manner. This strategy is challenged in crisis situations. But the brain is so powerful that it is basically capable to return regularly more or less quickly to the path of constant optimisation, so that there is nearly no problem that cannot be solved in the end.

But there are also reasons why the optimisation strategy or the whole system, respectively, reaches its limits – either temporarily or long-term – this topic will be broached by "The 4DI model" – see the appropriate chapter below.

Universal suitability probability evaluation

This section tries to explain why there is the talk of "suitability probability evaluation" and why the brain is classified as "suitability probability processor".

Term "evaluation" has to do with the following concepts or aspects of the brain or of the SPP model, respectively:

- **Emotivational signals** that are emitted by the autonomic interface, pain sensation, emotional processing and by the reward system. These signals are transferred to crucial parts of the brain, as the basal ganglia (multiplexer, affective goal decision making), the frontal cortex (associative control, value-guided decision making) and the association cortex, namely the entorhinal cortex (memory). This has been outlined in section "Homeo-stasis, pain, emotions and rewards". Term "emotivation" has been introduced for this class of phenomena in section "The emoti(onal-moti)vational system". Further concepts have been added in section "The attention assessment controller (AAC)". A crucial point here is that emotional and motivational signals can be emitted by different sources, but that all these signals, which represent any kind of purpose, are compatible in general, so that they can be compared and evaluated in concurrence to each other. This (brain-) universal aspect is the reason why "emotivation" has been introduced as abstract term for all kinds of purposeful signals.

- **Assessment,** as explained in the ACC section above. Each content in the associative memory is enriched with emotivational value patterns, i.e. patterns that represent autonomic states, pain, emotions or rewards. This takes primarily place via associations between informational patterns and emotivational values in the association cortex. Furthermore, the weights at all excitatory and inhibitory synapses in the whole brain can also be seen as values, referring to informational patterns which circulate through the nerve cords and synapses as excitations and which are transmitted and processed on the basis of values by this way. Value information which can be a basis for evaluation and filter processes has got an inherent ingredient of memorised information by these two mechanisms.

- **Attention control** and **decision making,** as explained in the AAC section above. This refers mainly to the value-guided decision making chain, leading from emotivational signals that are emitted by the emotional-motivational system over *value-guided goal decision making* in the prefrontal cortex and *motor decision making* in the anterior cingulate cortex to sensorimotor procedures that are consequently stimulated in motor cortex and sensory cortex. This comprises also value-guided associative processes in the prefrontal cortex and the declarative memory system in the sense of preparation processes, which do not always have to lead to a motor decision. And this refers also to the affective decision making chain through the basal ganglia over the anterior cingulate cortex to the sensorimotor control system; this is not attributed as value-guided, but the involved pattern transformation and decision making maps are based on synapses, and are thus in the broader sense also matter of evaluation.

The concept "suitability probability evaluation", as to be explained here, focuses on the direct attention control and decision making processes. This includes the value-guided decision making chain and the associative preparation processes in the prefrontal cortex and the declarative memory, while it excludes the affective goal decision making chain and values that are coded in synapses. However, the latter things are also matter of regulation and optimisation, but in a more indirect way via the former.

The crucial principle and common denominator for attention control and decision making is balancing of the brain processes for the sake of the individual and the species in the process of evolutionary competition. This is quiet similar to other control loops in nature and technology, and it takes place as weighting of actual values in comparison to set values. The set values are the needs which prevail currently or which are in standby (e.g. bodily integrity is always a standby need, sexual arousal can be). The actual values are the values that are achieved by the current activity process for the body or by the current preparation process for the resulting associations. Regulation is necessary when the results do not meet the demanded need satisfaction, so that the emotivational set value signals rise or turn negative instead of decreasing or turning positive – in the sense of aesthetic experience – as demanded.

A specialty of the brain control processes, compared to simple control loops, is, that multiple set values and potential (standby) set values are involved at the same time. The consequence is that this is not only a control loop for one parameter and set value, but many set values concur typically to one another. That's why brain control must have two stages – goal decision and solution pattern decision.

Bot stages are evaluations between concurrent patterns with value component. Making a decision for a goal means evaluation of the importance of all involved goals and find out which has currently the largest significance for the organism. The direct emotivational value signals of the involved needs are compared to each other in this case. A simple formula can be that the strongest need, which causes the largest emotivation, will win.

The winning goal from the first stage provides the set value for the next stage – the motor decision or the search for an adequate associative pattern. The actual value is now provided by the solution pattern, which is currently excited – either in the sensorimotor control system for the case of a motor decision or in the associative/declarative areas for the case of preparative reflection. Regulation means in these cases that the current solution pattern is continuously compared with other candidate patterns and that a better pattern is immediately selected if its value becomes the most promising in the current situation, in relation to the demanded set value.

It must be clear that external and internal information is always the other crucial factor for the pattern search and (pre-) selection processes – it would be useless to select patterns which promise great values but by matching to completely different physical or informational contexts; the informational part of the lookup and selection process or pre-selection process works via external and internal perception (sight, hearing, touch, kinaesthetic sense etc.); for the associative/preparative processes this might be linguistic processing due to hearing or vision, or it might only be the former train of thought, which recommends itself to being completed. But value comparison is always the most decisive final part in the decision chain, because it represents the purpose which is behind all these processes.

However, the goal decisions as well as the solution pattern decisions are processed via synaptic multilayer networks or decision making maps, respectively – what this means, is described in section "Excitation, inhibition, pattern transformation and circuits" in chapter "Neuro basics" and in Glimcher 2014 and references therein. The crucial and final part of these decisions involves always the emotivational values in a universal way. The result is that always the goal is selected, which may represent the biggest advantage for the individual in the current situation, and that the solution pattern is selected, which meets the current context at the best and which is accompanied by the best assessment value in relation to the demanded emotivational value.

The general criterion for this kind of pattern selection processes is always to find the patterns with the best suitability in relation to current needs and context. The sense is, that the goal must be suitable for improve the state of the individual, and the solution patterns must be well-suited for achieve or approach the goal. That's why it might not be wrong to entitle this principle as suitability probability processing, leading to the concept "suitability probability processor" (SPP) for the brain and for this part of the book.

In this connection, suitability *probability* means not that this is based on accurate data collection and mathematical statistics, but it means rather that it is "only" a kind of suitability probability *estimation* system.

The efficacy of this system should nevertheless not be underestimated, because it is target of constant optimisation and it is field-tested by the successful development of mammals and humans. The results might not always be very exact in the mathematical sense, but a great advantage should be that they are always totally suitable and up-to-date, because it is ensured that all important aspects and gained optimisations are definitely involved with no latency. This is a big pro in comparison to mathematical statistics which are often made available by effortful studies and often applied to contexts later on, which have been changed in between (note: this disadvantage might be settled in contemporary big-data solutions, which are increasingly able to integrate gain and application of data).

It is a central book of the SPP model that nothing happens in the brain and nothing is stored in the memory without regard to purposes, emotivational values and needs. But what about the steady perception of the environment? Per-ception seems often occur in general, without direct purpose and motivation. An individual is typically able to observe his environment and to record what happens around him, de-spite the fact that not all events are relevant for him. How might this be consistent to the book that nothing happens in the brain without purpose and value?
First, observation, although if widely comprehensive, is always a filtered process; a lot of irrelevant signals are put aside as noise and others are prioritised as important enough for recording. Second, there are many potential purposes and needs making constant observation sensible. Examples are the need for safety – causing the aspiration to detect ir-regularities as early as possible –, any

oriented action – requiring constant orientation –, or kinds of play instinct or artistic ambition – with unbiased marvel at everything. Third, if the goal of current activity is classified as profoundly important, perception works consequent-ly in a highly focussed way. The book is here that excitation patterns do not leave traces in the nervous system without purpose and without a dedicated need, even if it seems so on superficial consideration.

A difficulty is that there are many different types of emotional or motivational signals. There are at least the four general types – autonomic signals, pain sensation, emotional patterns and signals representing rewards – which have to be divided into a larger amount of more concrete types, as e.g. hunger, thirst, pressure pain, headache, any kind of feeling or even the desire for any concrete reward. This leads to a certain necessity for differentiation. See also section "Homeostasis, pain, emotions and rewards" in chapter "Neuro basics" and section "The emoti(onal-moti)vational system" above.

But there is also the insight that decision making on the one hand, so the attention part in the AAC concept, and assessment, so the assessment part in the AAC concept, are both working in a unified way, independent from the concrete goal and action strategy in each case. And these processes are well-suited for manage combined evaluations and decisions in a universal way, independent from the types of the involved emotions or motivations. Pain can be balanced in relation to hunger, and pain and hunger can be balanced in relation to a reward, as e.g. a mountain peak. This outstanding universal capability of the cerebral control system can only become identifiable by adequate levels of abstraction and generalisation. That's why describing the interface to the autonomic system, pain sensation, emotional processing and the reward system as generalised interface to an abstract "emotivational system", as done in section "The emoti(on-al-moti)vational system", is adequate, when focusing on the general characteristics of the nervous system.

Following this path, it is only consistent to combine emotional-motivational to the special term "emotivational", as well as emotion and motivation to the term "emotivation". Furthermore, "emotivational value" might be a sensible term for the currency which is paid in the brain in general and which makes the universal comparison of any purpose-related component of excitation patterns possible.

It can be the talk of an emotivational pattern which is generated by the emotional-motivational system, when a deficiency occurs in the autonomic regulation system or when a pain is perceived. For the content of the declarative memory, this leads consistently to the universal naming as "associative-emotivational patterns"; this refers to all kinds of brain areas that are involved into the associative processes, controlled by the associative/cognitive control circuit through the basal ganglia, and proclaims that the universal contents in these areas can always be entitled as associative-emotivational patterns, in the sense that these are informational associative patterns that are coloured by emotional experiences or emotivational values, respectively.

Needs and library of associative-emotivational patterns

For the further differentiation, though on the path of generalisation, it is necessary to bring term "need" more into focus. What are needs and why is this concept important here? Term "need" is just another, but rather more psychological term to specify and classify signals that are elicited by the emotional-motivational system. A need arises when a pain, autonomic deficiency or urge for a reward occurs which cannot immediately be satisfied. But it is a typical situation in daily life that a specific need occurs not segregated. It is rather normal that multiple needs concur to one another. The situation gets more complex when considering that there are, besides active needs, always also any number of standby needs that can potentially become active due to the further developments. This can be, for example, a need for inviolacy, which represents risks on the way to the demanded food that may satisfy a hunger need.

The result is that multiple needs, that are either active or potentially active, concur typically to each other. And it is exactly the decision for a goal, which is either processed as affective decision via the basal ganglia or as value-guided decision via the prefrontal cortex (or in coordinated manner by both instances), as described above in relation to the control levels (3), (4) and (5), which resolves this conflict. The constant work of this hybrid decision making apparatus, or of the attention part of the attention assessment controller, consists of constant balancing between needs that concur to each other in the particular context. Thus, the decision for a "goal", as described in some scientific papers, is at the same time the selection of a need that is allowed to turn into the prevailing need. This is the same process, but only described with different words.

The advantage of term "need" is that it opens another perspective of the goal decision making or attention control process, which refers more to the psychological dimension of the matter rather than only to neurophysiological processes. In the light of this perspective, emotivations represent also needful states and need satisfaction experiences rather than only excitation patterns and value components in decision making processes.

On a very abstract level, needs can be divided into basic

needs and higher needs. The relationship between needs and emotivations can be explained in different ways for both categories.

Let us start with **basic needs**. The initial purposeful component which takes effect in the nervous system is homeostatic regulation, i.e. the necessity to guarantee that certain biochemical preconditions for being able to survive are not violated. Besides this there are also sensations of pain and other alarms signalling that the body is in a bad state, which has to be avoided for the same purpose. This is the primary material for basic needs from which the human cannot escape (see also the explanations in the sections above).

For homeostasis and for the interfaces to the vegetative/autonomic nervous system, representing bodily states, it is known that this is bound into the emotional-motivational circuits via the hypothalamus area, and pain sensation is typically involved via the so-called thalamus. This is the way in which changes of bodily states or pain are incorporated into the equation, providing an aim for the neural processes and leading to a basic kind of emotivational colouring of the patterns which are circulating in the emotional-motivational system and adjacent networks. So it can be assumed that direct need satisfaction activity via control levels (3) and (4) always generates combined excitation patterns which involve sensorimotor parts, associative parts and basic emotivational parts. And it can be assumed that these excitation patterns can potentially be experienced (if occurring consciously), and that they might be stored in the cerebral cortex as memory content. An important conclusion should be that associative patterns, comprising microrepresentations of sensorimotor patterns, always include also the experience of emotivational gradients as intrinsic component, at least if acquired by need satisfaction activity for basic needs.

This is not different for **higher needs** or **"artificial**

needs". How higher/artificial needs can evolve as needs – in terms of becoming a goal or an emotivational stimulus – will be explained in the next section below. But a pre-stage is provided by pattern combination. Each type of perception and activity, as well as the processes of situational and creative preparation, as described in section "The control levels of the nervous system" above, always result in new combinations of associative patterns which evolve in the brain, and which are potentially experienced (if occurring consciously) and are potentially stored in the memory. When considering that the primary basis for each type of more complex activity and preparation process is always provided by basic experiences with basic emotivational components, as described in the paragraph above, the conclusion must be that emotivational components are also always included in complex pattern combinations.

The result is that the brain is an engine which is able to process, store, recall and combine associative-emotivational patterns, including associative microrepresentations of sensorimotor-emotivational patterns. That emotivations are always involved in the processes can be regarded as a general principle. The emotivational components might be stronger or weaker, but there are no associative patterns without emotivational components, just as there is no perception without associative or sensorimotor activity. Thus, the long-term memory can be seen as a huge library of associative-emotivational patterns, which is constantly fed and managed by action and thinking.

Higher needs

It is clear so far, that the brain works primarily on the basis of basic needs and that lots of need satisfaction incidents are stored in the memory. Continuous recombination processes, which occur in the preparation mode of the attention assessment controller (AAC), result furthermore in steadily growing combinations of associative-emotivational patterns in the pattern library, namely in the declarative memory. These composite patterns include complex combinations of emotivational colourings. But how can higher needs evolve from this?

The book here is that the neural patterns, which are stored in the memory, have another important function – they are also the blueprints for further needs, to be developed additionally. Higher needs might be produced via two stages – firstly by growing as combined associative-emotivational patterns in the memory and then, later on, by some of the results becoming prevailing needs for a while.

There might be three mechanisms to create higher needs and evaluation levels from the basic processes and the resulting emotivational value patterns:
- emotivational dependency chains,
- emotivational stimulus modulation and
- artificial needs.

Let us continue with explanations of these concepts.

Emotivational dependency chains

Emotivational dependency chains can only explain how complex emotivations can develop in the long-term memory, along with the combination process of associative patterns, and how they can be involved in the evaluation and decision making processes. But they cannot explain how a higher need can evolve as an emotivational stimulus. This is rather a matter for the other two mechanisms described below.

For *higher* needs, it might always be possible to describe

a dependency chain from basic needs and emotivations. Self-actualisation or art could, among other things, be motivated by appreciation, leading to social needs, better opportunities for safety, and fulfilment of basic needs. Realising this chain of dependencies via the mixture of involved emotivational components, right down to the purest parts, in a complex pattern, might be part of the truth. The evaluation process might be able to calculate all these components in the equation very quickly.

More and more complex dependency chains might develop, along with the combination of associative-emotivational patterns, from basic behaviours to more and more sophisticated attitudes.

Emotivational stimulus modulation

Homeostatic regulation and the sensation of pain might never really be totally quiet. So it could be that weak basic deficiencies are at least always signalled from the hypothalamus or other sources to the basal ganglia and the attention assessment controller. Therefore it could be a regular mechanism, that these signals are exploited to modulate a remembered emotivational component to these, and in this way turn the latter into an emotivational stimulus, also dominating for a while in relation to the underlying basic need.

Any kind of remembered combination of emotivations, stored in the memory in the past, can turn into an emotivational stimulus through this mechanism. This is true for basic emotivations, special emotions or feelings and also for any kind of complex emotivations developed by a pattern combination in the past. The recall of a memorised basic need satisfaction experience as an emotivational stimulus might be carried out in correlation with a real needful state of the body, matching its pattern at least partially, but not necessarily having to.

Stimulus modulation could be a way in which emotivational dependency chains or any other complex pattern combinations can turn into an emotivational stimulus. The idea behind this is that any good memory, which may have occurred in the past, can at any time later on become a demand, and that a weak basic deficiency is adopted to apply the pursuit to another idea. That a memory can become a stimulus again is only natural because it is just a remembered need satisfaction incident or a combination of such incidents.

Higher/artificial/virtual needs

The idea of artificial or virtual needs is that a general pursuit of good feelings results from basic experiences, and that this is maintained and developed in any direction. The basic experience that one can come into difficult situations and states, or that pure deficiencies can happen, causes a general ambition to oppose this kind of occurrence. And this is a reason why the more complex emotivations, which arise in relation to the more complex informational contents of the brain, can also result in the development of higher or artificial needs.

Complex emotivations and artificial needs can to a great extent be disconnected from the rationale of pure states of deficiency. A world of composite emotivations and higher needs is developed, which becomes more or less independent from the gradients of pure states of deficiency from which it originated. This world of feelings and ambitions can grow in any direction and it can become broadly independent from the basic states. A remaining restriction might be that from time to time it is validated at least partially in being able to deliver results which can help to manage real life (that is, basic needs).

The world of composite emotivations and higher needs might, among other things, develop with the help of the mesolimbic dopamine system (the reward system). This mechanism works in a way so that complex emotivations (good memories) can become stimuli, bound into the evaluation processes of the AAC circuits, just as this occurs with signals from homeostatic regulation (it is not by chance that the mesolimbic dopamine system is an integrative part of the BG/ AAC system). The result is that evaluation processes, which occur in relation to higher or artificial needs, can take effect via the same basic mechanisms as occur in relation to basic deficiencies. This kind of generalised working principle makes it possible that the capabilities of the brain can be optimised in an integrated manner for all kinds of needs. The mechanism of how self-regulation takes control over behaviour, thinking and memory is always the same.

The book is that artificial needs always apply to complex or composite emotivations. So it is possible for value sys-tems to be established which are far above basic rationales. Of course, these are initially caused by basic needs in terms of having the same roots, but they are completely decoupled from this by emotivational abstraction. So it can be possible to develop needs and emotions which are associated with a beautiful-looking garden, a work of art, a motorcycle, a family celebration etc. Cultural achievements are the visible phenomena for the worlds of composite emotivations and artificial needs that are developing in the minds of the peo-ple.

Artificial needs are not necessarily assigned to sufficient behaviours, elation and great cultural achievements. Counterproductive behaviours or deplorable developments can also result from this ability, such as, for example, addictions or inhuman attitudes.

In the contemporary world, it should be the rule that motivations are driven by remembered needs and especially by complex compilations of needs, dependency chains or artificial needs; that they are driven by real pain should be the exception, at least for normal life under good conditions.

It is also important to mention that memories not only represent positive incidents and stories of success, but also failures and frustrations. Just as inhibition is part of the truth at the synapse level, so are negative experiences of any kind part of the truth at the sensorimotor-emotivational and associative-emotivational levels. The road to successful behaviours is always paved with numerous unsuccessful attempts. These are also stored in the pool of experiences. And they are important as differential anti-patterns in the evaluation process, signalling both the potential impact and the behavioural variants to be avoided, as far as possible. They are essential to always obtain an internal stress field between suitable and unsuitable answers to complex situations, representing the real impact of the matter in a broadly sufficient manner, since it is better to be able to fight this out internally prior to always generating bad experiences again and again. So it is an important aspect of the evaluation process that the stress field between positive and negative experiences is aligned to the prevailing pursuits.

The world of artificial needs, which can be cultivated by humans to an extent which has never be seen before, should be seen as the most beneficial feature, which makes humanity the most successful species on earth. The ability to develop intensive social communication and cooperation is the other precondition for the profound success of the humanity, as a matter of course, at which both advantages are to be seen as closely intertwined.

Needs and suitability probability evaluation

Above it has been described of how evaluation and decision making processes take place on the basis of emotivational excitation patterns. This is very abstract, and another sight to this matter can be to break this type of processes down to different types of needs.

The question is: how occur evaluation and decision making processes in relation to different needs? This question can be asked in relation to a basic or a higher need, as well as in relation to activity, accompanied by situational preparation, or in relation to creative preparation.

The following variables are always involved by the evaluation processes that are ongoing in the attention assessment controller:

(a) Basic needs: The state of the body with all its parameters, e.g. biochemical conditions or states of pain, incorporated via the hypothalamus or other basic emotivational interfaces.

(b) Higher needs, evolving in a seemingly detached way from basic needs, and preferably if basic needs do not prevail.

(c) The environmental context, incorporated by perception and sensation.

(d) Real activity, controlled by the sensory and motor cortex and going on with the help of lower regions of the nervous system as, for example, the basal ganglia, the cerebellum etc. (control level three), by being supported by an appropriate thread of situational preparation (control level four).

(e) Some threads of creative preparation and linguistic simulation (control level five).

Each of these variables can become quiet for a certain time – (c) and (d) by sleep, (a) if all basic needs are fulfilled and no pain prevails, (b) if basic needs are active, (e) if current activity requires full attention. But all these variables are typically in the equation.

Let us now try to find some answers.

(a,c,d) How is the pattern search ongoing in relation to a basic need and real activity, along with situational preparation?

Let's start with the motor decision topic. A basic need has been selected as goal, the current emotivational stimulus is provided by this prevailing basic need. It is emitted via the interface of the hypothalamus to the basal ganglia and the attention assessment controller. The emotivational set value is a positive (satisfactory) gradient for this need.

The current context is the physical environment of the person. The informational aim is to carry out real action in a successful manner, due to the qualified perception of the environmental conditions.

The objective is to obtain patterns from the library which are able to meet the informational aim, and the emotivational set value, at the same time. This is a difficult task, since each situation regarding both the informational and the emotivational part should be more or less different to the situations managed in the past. No pattern will match exactly.

Situational preparation is carried out to play through the options available. It should start from the perceptions and actions which are currently ongoing with the help of habitual and voluntary components. Preparation means that the sensorimotor activity procedures (procedural) are accompanied by corresponding processes in the associative parts of the cortex (declarative). This means that associations are activated, which include microrepresentations of the ongoing sensorimotor-emotivational sequences and which involve also adjacent or similar associative patterns. The weights of these excitations are typically slightly varied, what might lead to a play through different alternative abilities for continue the interaction process and finally to a motor decision influence for the activity sequence, which is executed really.

But what are the right or best alternatives in the sense of evolutionary competition? Answering this question is the task of emotivational evaluation and decision making. The emotivational components of the patterns, examined by the preparation process, are evaluated in relation to the emotivational set value. This should lead to the pattern with the highest probability of satisfying the prevailing need in relation to the current environmental situation. Another aspect is that patterns are classified as to be avoided which represent negative experiences in relation to a similar situation, signalled by negative emotivations. This is especially important if a kind of trial-and-error strategy becomes necessary, caused by a lack of positive patterns.

So it is a filter process, which is constantly ongoing in our brain in a highly efficient manner and which is a great solution under the circumstances of ever-changing environmental conditions. Emotivational evaluation and decision making provides the ability to always find the pattern which is most probably able to solve the situation, comprising inner and outer conditions. This works independently of whether the resulting suitability probability, measured in absolute terms, is really high or still very low – it works in a comparative way, by simply selecting the best pattern available in a quick and efficient manner, or at least by avoiding patterns, representing unsuitable reactions, in a reliable manner (see also Singer 2004). That is why we can speak about the brain as a suitability probability processor, at which this means suitability probability estimation rather than exact statistical calculation, as already stated above.

But the examined case is quite simple, providing an explanation of the basic principle. However, the reality is much more complicated. A variety of demands and needs must always be considered which might conflict with each another.

Now it is time to consider the goal decision topic additionally. Let us take an example with two needs – thirst and avoiding pain. Let us assume a person who is actually driven by the former need, thirst. And the way to the water seems to be not far away – a creek with clear water is seen in a valley, just a few hundred metres ahead. But there are different options: a direct way, but leading over a cliff, and other ways, which are longer and less clear-cut. The thirst assessment will certainly vote for the short way. But ignoring the risks of the cliff is not really a good option, and taking this fact into account makes it clear that thinking that the evaluation process is driven by a single need, or only the need which is currently prevailing, must be insufficient. Instead the process incorporates multiple needs, including, certainly, all the needs that are currently active, but besides this, also those which can potentially become active and which might be called standby needs.

What happens in the evaluation and decision making process is that all types of emotivational colours are included in the equation. All the involved components are able to cause feelings reproducing the actual need-state-change gradients incorporated during the creation of the patterns (during the primary experience). That is why all human needs which could be affected by the selected patterns are implicitly evaluated and calculated concurrently with each other.

The thirsty person might primarily evaluate the direct way, but by taking into account the fact that climbing down the cliff includes certain risks of serious injury, i.e. by anticipating the risked pain. This includes the goal decision making instances in the basal ganglia and frontal cortex, which compare emotivational values of the safety need and thirst to one another. The goal decision making system and the motor decision making system may cooperate in this case, leading to the result that a detour to the water is selected. This might be more complicated than the direct way over the cliff, but both needs are involved and satisfied in the end, the safety need for integrity of the body and the thirst.

At least two further possibilities should be examined if this kind of useful anticipation (preparation) is missed and the direct way over the cliff is chosen, despite the risks. Switching the prevailing impetus from thirst to the objective to avoid pain could become imperative when the person finds himself in difficulties in the abyss which has to be climbed down. Thirst will become a secondary issue for a while and might reoccur when the difficulties with the cliff are overcome.

The third possibility to be mentioned, as is clear so far, is that there is a really bad experience with the cliff, and there is an injury.

Negative experiences will become especially important if there is no solution which is more or less obvious. This can happen, if thirst meets a situation where no creek is seen a few hundred metres ahead and where no other source of water is visible. The patterns that represent positive values are scarce goods in this case, the positive signals are relatively weak, and explorative strategies will become necessary, leading to inventive search processes and behaviours. But the library always provides a rich set of anti-patterns, issuing negative emotivational signals revealing possible responses which have to be avoided. So they are a great support for the decision making processes and the further search for the desired good.

Therefore it can be concluded that this is a mechanism by which the brain is able to meet any situation with great effectiveness, scaling up from those where well trained, optimised behaviours are available, to scenarios referring to a vast lack of positive suitable responses.

(b,e) How does the evaluation process work in relation to higher needs and creative preparation?

The principles are the same as in relation to basic needs, real activity and situational preparation.

The differences are:

- That associative-emotivational patterns are searched in the associative part of the cortex and in the declarative part of the memory system, respectively, rather than sensorimotor-emotivational patterns in the procedural sensorimotor part of the cortex. In this case, the latter are only represented as microrepresentations in specific areas of the association cortex. This point is also already true for the situational preparation processes, which are mentioned in the descriptions for case (a,c,d) above.
- That a higher need is always based on more or less complex combinations of pure needs and basic experiences.
- That a higher need is never caused as a pure signal of pain, but rather as an artificial/virtual need or dependency chain need, becoming a prevailing need by emotivational stimulus modulation or via the reward system, or any other sophisticated mechanism.
- That the pattern search and evaluation process is more or less decoupled from the real context and activity of the person.

A real activity program should be ongoing in parallel, maintained by threads of situational preparation as described above; this could be walking or keeping a sitting position on an office chair. But there are still resources for the ability to also concentrate on creative preparation, to a certain extent. The aim could be, for example, to solve a problem or to get an idea of how a project could be brought a step forward, which occurs in relation to a higher demand.

The associative-emotivational patterns demanded should not necessarily refer to the environmental context in this case, but to the context of the problem to be solved, or the project which is targeted. So this process can evolve in relation to any context – even far away in time or space, fictional, scientific, religious, in obscurity, or anything else. And it can occur quickly or over a long period or it can flare up again and again. So there is seemingly no restriction or limitation in general. The actual restriction lies in the demand or motivation which is behind the pattern evaluation process. And it is typically necessary to ensure that real activity is involved in the end, for example by using a pen or a computer to write down the results, or by scheduling a meeting in order to have discussions about the matter. Real activity and communication might also become necessary due to the finding that new neural patterns should be established in the library or in the long-term memory, respectively, in terms of knowledge acquisition (associative-declarative) or training (sensorimotor-procedural).

Suitability probability evaluation and evolution

As a summary we can state that suitability probability computation is constantly ongoing in the brain as a competition between neural patterns in relation to real activity or any distant contexts in terms of time, space or closeness to reality.

The constantly ongoing evaluation processes have different results:

- Goal decisions: constant competition and balancing between different needs, which are represented by characteristic emotivational patterns with different strengths and gradients, but which are compatible in general so that they can be compared to each other.
- Solution pattern decisions: evaluation, filtering and selection of patterns which are sufficient in relation to the external context and to the goals from the point above, in terms of prevailing needs. There are two major types of processes and corresponding types of solution patterns: situational evaluation and motor decision making processes, demanding sensorimotor-emotivational solution patterns (in the sense of external interaction), on the one hand, and creative associative evaluation and decision making processes, demanding associative-emotivational solution patterns (in the sense of internal reflection), on the other hand.
- Creation of new pattern combinations via multiple processes or threads of preparation and activity control, occurring simultaneously or in parallel under control of the AAC, and by emergence resulting from the ability to mix or combine excitation patterns between the different threads.
- Inscription of some of the patterns as experiences into the episodic and long-term memory (declarative and procedural), resulting in constant maintenance and enrichment of the libraries of sensorimotor-emotivational and associative-emotivational patterns. This includes to a great extent also patterns representing perception incidents, since situational control processes include constant perception over all senses.
- Some of the patterns available in the library can be re-

called to be made again as positive experiences, in this
way becoming prevailing artificial needs.

- Negative experiences are the other side of the same
  coin, important as anti-patterns, representing dangers
  and risks in the internal evaluation process, and supporting exploratory trial-and-error strategies if a lack of positive patterns occurs in any situation.

The actual purpose of all these processes is the general
ability of the individual to regulate his affairs in a sufficient
manner according to the demands of the evolution. This is
primarily incorporated by real closed control loops, as homeostasis and other basic mechanisms, but it leads to much
more complex processes of reproducing internal images of
the external world and exploiting this to apply sufficient behaviours. It is about the continuation of the evolution by
different means, supplementing the competition of species
in physical and physiological spheres, increasingly, also, by
competition between neural patterns.

The general suitability probability of these processes depends on the following factors:

- The complexity of the interconnections in the nervous
  system. The more types of perceptions that can be involved in a coherent pattern, the better the chances that
  a suitable picture of real life is reproduced internally. "In
  palpation there is first the shaping of the hand for grasping an object, and secondly the moving of the hand over
  the surface of the object in an active exploration. In this
  way cutaneous sensing leads to feature detection that
  matches the visual feature detection in the inferotemporal lobe" (see Popper and Eccles 2006, 260, Chapter E2:
  "Conscious Perception", section 9.: "Cutaneous Perception (Somaesthesis)", sub-section 9.3.: "Secondary
  and Tertiary Sensory Areas"). The attempt to name the
  object and to vocalise this develops another dimension.

"Particular importance is attached to Brodmann's areas 39 and 40, which came very late in evolution, being barely recognizable in nonhuman primates. These are the areas specifically concerned in cross-modal associations, that is associations from one sensory input, say touch, to another, say vision [...]. It is postulated that language comes when you have the association between objects that you feel and objects that you see, and which you then name. [...]. Language provides the means of representing objects abstractly and for manipulating them hypothetically in one's mind." (see Popper and Eccles 2006, 295f.). Incorporating human speech into the system of interconnections has two aspects in this regard – it is another ability of the brain to develop interconnection complexity and it opens the social dimension – see the point after the next point for details.

- The above factor should be seen as an informational quality. Another important factor is the emotivational aspect. The better the internal stress fields, provided by the emotivational components of the patterns, are able to meet real life conditions and the impact of human activity, the better the chances to cope with the demands of evolutionary competition.
- The complexity of social interconnections. The ability to express and communicate perceptions and to maintain cooperative activity opens the way to a dimension of the general suitability probability which distinguishes humans from mammals, to a great extent. It is not necessary to explain the great potential of this dimension here since it is obvious when you see society's achievements. Communication involves both the informational and emotivational components of neural patterns. The social aspect, in particular, also provides a great dimension to enrich the neural pattern library with suitable

content, becoming more and more important due to scientific-technical progress and globalisation.

The two types of consciousness

A question which is always especially interesting, is: "What is consciousness?" There are lots of answers from different points of view in philosophy, neuro sciences, social sciences and other fields. The SPP model tries to contribute another perspective.

The answer has two aspects: There are two types of consciousness – **informational consciousness** (informational aspect) and **emotivational consciousness** (emotivational aspect). Both refer to the same procedures – the associative-emotivational patterns processed currently in the brain - but to different aspects of this. Consciousness is simply a question of quantity in relation to both aspects.

The evaluation processes, which are constantly continuing in the nervous system, involve many areas of the cerebrum. Consciousness is the ability to notice these processes in terms of inner experience. But large parts of the processes are too weak or too volatile to be noticed. A pattern only becomes conscious if a certain quantity is reached for the product of time and significance.

For the same sequence of excitation patterns the question of whether it exceeds the threshold to consciousness is answered independently for the informational and emotivational aspect. Both aspects can become conscious, or one of these, or neither.

Large parts of the processes of evaluation and assessment that are constantly ongoing in our brain have to be attributed to the latter case – it is always ongoing in a completely unconscious manner. Only in some exceptional cases,

Table 1 – Cases in relation to informational and emotivational consciousness

c — case
ic — informational consciousness
ec — emotivational consciousness

c	ic	ec	phrases, description
1	No	No	Subconscious mode, subconscious mind. This is the actual innate demand of the brain. The majority of responses are produced with high efficiency in this mode.
2	Yes	No	Sober or academic mode, sober/unemotional way of thinking. The informational part is processed with a certain significance, but emotivations are mainly involved as unconscious filter mechanisms, rather than as significant perceptions. This may be accompanied by a remaining abstract gut feeling, representing the fact that emotivations are involved behind the scenes.
3	No	Yes	Emotional mode, emotionality. The emotivations are overwhelming and it is difficult to uncover the actual causes, reasons and interdependencies in rational manner.
4	Yes	Yes	Committed mode, committed way of thinking. Reason is unified with feeling to a certain extent. The balance between both parts can vary over a large range.

if suitable responses to a situation or problem cannot be produced quickly and efficiently, then consciousness arises. So it is actually a general immanent pursuit of the brain to solve the demand for suitable patterns as quickly and as inconspicuously as possible. Consciousness is only the capacity to head towards residual problems which cannot be solved as discretely as is actually demanded.

Based on the book that this happens independently for the informational and emotivational part, four cases can occur, as described in table 1.

The common understanding of *emotions* (not emotivations in this case) might correlate with the latter two cases to a great extent; rationality is preferably assigned to case 2 and mysteries, magic and psychoanalysis to the subconscious (case 1). But this differentiation is apparently unsubstantiated. Emotions and rationality, as well as unconscious and conscious, are rather just aspects of the same great achievement of our nervous system (see also Singer 2004), called the *suitability probability processor* in this book. This engine is able to exploit the four aspects with the greatest adaptability ever imaginable – ranging from fast responses, near to the physiological limitations, to extremely slow thinking and problem solving, and from well-trained behaviours to the complete lack of positive patterns (see also Singer 2004, Kahneman 2012 and section "Kahneman, Daniel: Thinking, Fast and Slow" in chapter "Psychology" in Part 2).

Conscious experiences

The question of consciousness is also discussed as an "explanatory gap" between subjective experiences and their objective neural basis. In this regard, Block (2009), and references therein, wrote (1113):

"Phenomenal consciousness is 'what it is like' to have an experience (Nagel, 1974). Any discussion of the physical basis of phenomenal consciousness (henceforth just consciousness) has to acknowledge the 'explanatory gap' (Nagel, 1974; Levine, 1983): nothing that we now know, indeed nothing that we have been able to hypothesize or even fantasize, gives us an understanding of why the neural basis of the experience of green that I now have when I look at my screen saver is the neural basis of *that* experience as opposed to *another* experience or no experience at all. Nagel puts the point in terms of the distinction between subjectivity and objectivity: the experience of green is a subjective state, but brain states are objective, and we do not understand how a subjective state could *be* an objective state or even how a subjective state could be *based in* an objective state. The problem of closing the explanatory gap (the "Hard Problem" as Chalmers, 1996, calls it) has four important aspects: (1) we do not see a hint of a solution; (2) we have no good argument that there is no solution that another kind of being could grasp or that we may be able to grasp at a later date (but see McGinn, 1991); so (3) the explanatory gap is not intrinsic to consciousness; and (4) most importantly for current purposes, recognizing the first three points requires no special theory of consciousness. All scientifically oriented accounts should agree that consciousness is in some sense based in the brain; once this fact is accepted, the problem arises of why the brain basis of this experience is the basis of this one rather than another one or none, and it becomes obvious that nothing now known gives a hint of an explanation."

But, according to the SPP model, as described in this book, there is no explanatory gap.

But before an answer can be provided of how the explanatory gap can be seen as closed, it is necessary to explain that the question of phenomenal consciousness, as it is described above, has at least two aspects, which must be distinguished consistently. An "experience of green", to take the example from above, can be regarded in the differential sense, in terms of how green can be distinguished from other colours and shades of green, and in the literal sense, in terms of why this is expressed in a specific way, e.g. by the term "green". A "'what is it like'" answer can only be given in the differential sense, but not in the literal sense.

The literal sense would better fit a question like "Why am I used to describe this experience of green that I am able to precisely distinguish from the experience of red, as 'green'?" The answer is, it is why this is a practice or a convention, but not why it is determined in any way. The experience of green could also be described by the term "verde" (Italian, Spanish) or "grün" (German) or "jokuta" (fictitious). So each kind of experience can be associated with any linguistic construct in our neural networks – this is arbitrary, it is never determined in any way and there is an infinite creative leeway provided in this regard.

It can be concluded that the explanatory gap cannot be meant in the literal sense, because if this were the case, it would be a nonsense problem. This is also valid for any other forms of expression, not only language, as well as for any other cross-modal interconnections. The nervous system works in differential manner, and specific descriptions and linguistic expressions are only attached when it is attempted to communicate the phenomena. But how it is described and which words are used occurs always by chance rather than in a determined way. Natural laws can explain why language is used, but never which words are associated with which phenomena.

Concentrating on the differential sense of the explanatory gap, which is the only sense that is not nonsense, it can be stated that there is no longer an explanatory gap along with the SPP model. Let us summarise the concept of consciousness and follow this up by closing the (differential aspect of the) explanatory gap.

According to the SPP model, consciousness (of an organism or any other system) is based on the following achievements:

(a) The organism is equipped with any kind of self-regulation which brings a purpose or a relevance component into physiological and neural processes. This aspect is represented by emotivational value information in large parts of the neural processes and excitation patterns.

(b) Any kind of complex information processing ability is developed in terms of cross-modal associative-emotivational patterns and informational resolution ability according to the SPP model, or in terms of integration and the differentiation of information according to the integrated information theory of consciousness (IIT; Tononi 2012).

The fact that self-regulation is existent and emotivational value information is incorporated (a) leads to the demand that alternative patterns are always evaluated in the light of a purpose. This is the actual reason why there is a "difference that makes a difference" (Tononi 2012, 294, originally Bateson 1972).

(c) Preparation processes, according to control levels four and five of the SPP model, arise to the extent that they are able to occupy a major part of the neural processes concurrent with activity control demands (control level three) for significant time periods, with significant intensity. They are able to prevail in a decisive part of the neural information processing engine as constantly ongoing processes.

The fact that self-regulation is always incorporated in the form of emotivational value information (a, b) has the consequence that information processing is perceived as relevant for the organism itself. The longer lasting or more intensive parts of the preparative evaluation processes are what is perceived as conscious experiences (or as qualia, according to Tononi 2012).

So **consciousness is information enlightened by emotivational evaluation**.

How can the explanatory gap be closed?

For the ability to understand the relationship between the physical or objective aspect – nervous system, brain, synapses, networks of neural links, neural excitation patterns – and subjective informational processes – such as experience, mental states, decision-making, thinking etc. – it can be assumed that this is like a relationship between medium and content, or between medium and information, that is projected on its behalf. This is just as a CD is a medium for storing and playing music, the lens of a camera is a medium for obtaining a picture from a view, or the mirrors of a DLP projector are a medium for projecting a picture on a wall. The medium is made of matter in all these cases, but the processed information can neither be reduced to it, nor can it be resolved in the medium. One has to go to the focal plane(s) to understand what happens – e.g. the acoustic waves in the air, the film or sensor in the camera or the projection wall which is illuminated by the projector.

The focal planes of the nervous system can be described as follows: These are the environment, the body and the self-regulatory evaluation process. The environment is involved as a source of boundary conditions and as a target of activity. The body is involved as a source of basic needs and a target of self-regulation. The third focal plane occurs at the interface between the environment and the organism, which is provided by the evaluation process and the emotivational components of the neural patterns, or the filter processes of the attention assessment controller, respectively. So let us call this the evaluative focal plane. Brain processes would be completely indifferent in relation to any information, or the brain would not exist at all if this kind of evaluative focal plane were not available as a client.

Turning to the question "What is experience?", it is important to consider that the neural processes, as discussed here, are always ongoing in an unconscious and a conscious manner – see also the descriptions in the section on "The two types of consciousness" above. The more simple and basic part typically occurs as unconscious, which should also be seen as the original part, established primarily in phylogenetic history.

Consciousness/awareness, if it occurs, might particularly be devoted to the following boundary conditions (see also Singer 2004)[1]:
- Sensation of pain in terms of an alarm, forcing the brain to focus on an urgent matter for a while.
- Long-term solution processes, occurring because of

---

[1] In particular the descriptions in section "Conscious and unconscious processes" ("Bewußte und unbewußte Prozesse").

critical states and risks which cannot be solved immediately due to the lack of positive patterns, or because of problems which develop in a gradual manner in relation to solution approaches which are highly equivalent, or because of artificial needs which develop concurrently with one another.

- Subject matters that can be communicated in the context of problem solution and risk management at a social level.

These boundary conditions, particularly in relation to the latter two aspects, can only lead to consciousness

- if the information which has to be differentiated and integrated exceeds a certain level of complexity, according to Tononi 2012, and
- if the variety of evaluative requirements, as a specific kind of information, also exceeds a certain level of complexity and
- if the solution process exceeds a certain time threshold.

In this sense, **conscious experiences are the conscious part of the processes occurring on the evaluative focal plane**, notably the part which can be made a matter of communication (of any kind). That Nagel has an experience of green when he looks at his screen saver is just a direct projection to his attention assessment controller. "The neural basis of the experience of green that I now have when I look at my screen saver is the neural basis of that experience as opposed to another experience or no experience at all" (Block 2009, 1113) because the "difference that makes a difference" (Tononi 2012, 294, originally Bateson 1972) is provided by the evaluative demands which are incorporated into the neural processes by the attention assessment controller, which in this case means that Nagel is able to recognise whether his computer is in idle mode or not. Why? Because it is relevant for his demand to use the computer as a tool ... which is relevant for his work ... which is relevant for his life.

Much better examples might be provided by beautiful flowers or the magnificent plumage of birds, where small shades of colour can have great relevance. This leads from the pollination of plants by insects and the courtship of animals to graphic designers, who use computers for their work for the advertising industry (for example), and who will need 16 million colours rather than two, which might be sufficient for Nagel in the case above. This has further to do with the assumption that there is a general pursuit of aesthetics – more details see below (at the end of this section).

As a summary, we can assert that experiences and subjective states are identical with the suitability probability evaluation processes that occur on the evaluative focal plane. The sensory organs and perception networks of the nervous system are developed in such a way that they present the information to this focal plane in a differentiated and integrated manner, which is optimised for the purpose of suitable evaluation, filtering and decision making under a number of different circumstances.

But it is important to note that the brain is definitely not a target tracking system or a self-regulation system in a narrower sense. This follows from the following facts:

- The development of artificial or higher needs of any type plays a crucial role, as explained in the book over several sections. They can develop in any direction and do not necessarily have to follow an (easily) determinable purpose.
- Preparation, especially of the creative type, is a remarkable achievement of the human brain, resulting in unlimited capabilities for context exploration and a recombination of associative-emotivational patterns, as explained in the section "Needs and suitability probability evaluation" for cases (b,e). See also the descrip-

tions around concept "integrative encoding" in chapter "Emotion, motivation and memory" in Part 2.

These achievements make the brain a medium for creativity, and the evaluative focal plane supports direct and hypothetical projection "in relation to any context – even far away in time or space, fictional, scientific, religious, in obscurity, or anything else" (see the section "Needs and suitability probability evaluation" above). This can lead to results which do not occur on the other focal planes – the environment, the body –, either not at the same time or not at all. This is also a strong indication for not talking here about a reductionist approach.

Above it was asserted that there might be a general pursuit of aesthetics. What was meant by this statement? The brain does not only follow narrow evolutionary purposes. A crucial point is rather that the brain processes, although if controlled via multifarious emotivational patterns and concurrent needs, as described above, tend to follow a general principle, which can be attributed as "aesthetics" (note: this is only one of many abilities to name this). It is clear that the ultimate principle of brain control is provided by the evolutionary forces which are incorporated via basic needs. But it would not be sufficient to stay on this perception, at least not for humans. With the development of higher/artificial needs, which are directed towards rewards and positive experiences rather than avoidance and defence of pain and frustration, the brain tends more to become a producer of elations and aesthetic experiences than a stupid fighter against deprivation. This topic is outlined in more detail below in section "Dynamics of the need hierarchy".

Individual and social consciousness

Let us recapitalise consciousness of the individual.

In the section "Conscious experiences" above, consciousness was described as information enlightened by emotivational evaluation. It was explained that information from the environment and from the body is projected towards an evaluative focal plane in the cortex, which is a central part of the attention assessment controller (AAC), and that pattern evaluation and selection processes occur in an unconscious and conscious manner on this focal plane; the conscious part of these processes is what can be called conscious experience or qualia, respectively.

One crucial point is that a purposeful component is incorporated into the equation via homeostasis or basic needs, and higher or artificial needs. What happens in the AAC is that the informational parts of the experiences, also called sensorimotor patterns and associative patterns, are projected towards emotivational value components. Experience occurs exactly at the interface between the (informational) sensorimotor and associative components of the patterns on the one hand and the emotivational components of the same patterns on the other hand.

This is true for current real-time excitation patterns, occurring constantly in the brain in any form, as well as for representations of these patterns in the memory. The projection of information towards emotivation occurs primarily in the attention control or decision making processes, respectively, which are controlled by the AAC in activity and preparation mode, and which control behaviour and thinking in the sense of control levels three to five.

So far to the current real-time excitations. But projection of information towards emotivation performs also via the detour over memory content. The assessment of information by emotivational values is the major principle of memorizing processes in the brain. The result is memorised information with value component, which can be involved in further decision making processes at any time later on, and which is selected in the sense of solution patterns for any type of problems, which have to be solved by the brain and its owner. By this way, memories and their ability for being evaluated and reinstated are the most valuable resource for the primary attention control and decision making processes.

A question, which arises here, is: Conscious experience seems actually and primarily (on a naive view) to be a kind of perception process, as e.g. "the experience of green" – where is this aspect represented by the explanations of consciousness from above? Perception is always involved as informational filter for the pre-selection of memorised associative patterns and of sensorimotor patterns (or microrepresentations of the latter). Thus, perception is processed as search specification for potential solution patterns – either in the sense of sensorimotor activity or in the sense of associative preparation. Perception patterns are always gained by action – e.g. vision occurs via skeletomotor control – and they are always processed for the purpose of action or for the preparation of action. And all these processes are controlled by emotivational evaluation in the end. Hence, the answer is: Perception is never processed per se, but rather always in the context of action or potential action, and, to complete the story, action itself is never processed per se, but rather always in the context of purposes, so of emotivations and needs.

That perception, e.g. of a colour like green, may occur independently – with no intended action and with no purpose –, is most likely a rather naive view. Perception is always embedded into a purposeful background – reaching from perception as active target oriented process with motor control (*sensori*motor) over impulsive reactions or gain of preparatory associations in the context of prevailing needs to the constant evaluation of the environment in relation to potential or standby needs, which are always lurking behind the scenes.

However, the excitations which flow through the brain in the sense of perception, recall of memories, goal decision and solution pattern selection (activity or associative), are the processes which turn potentially into conscious experiences. These excitations have this potential because purpose is involved as emotivational component, so as current set values or needs, respectively, in concurrence to each other and as specification for assessment values of memorised solution patterns that have to be selected. These excitations occur preferably and to the greatest extent quick and efficient in unconscious manner, but they may always also turn into conscious state if deficiencies, complications or conflicts occur that cannot be solved as quick and efficient as usual. In this sense, consciousness is just a kind of quantification of subconscious processes, and unconscious processes are the raw material for consciousness, leading to the consequence that both parts are a unity with gliding transition between two major states. In this sense, *consciousness is information enlightened by emotivational evaluation*, as already stated in the last section above.

The purposeful component, which is provided by the pursuit to be successful in the process of evolutionary competition, is the only reason why experiences can occur and why they are able to reflect external phenomena with a certain precision. This is the only reason why there are ongoing informational transformation processes, which are a matter of constant optimisation, in the sense that patterns are resolved as being suitable on the focal planes in the different projection spaces. There would only be informational chaos if no optimisation had been ordered by evolutionary pressure for adaptation.

Three basic projection spaces are involved in these processes, with an explicit focal plane for each projection space, where the patterns involved can be resolved in a differentiated, integrated and exclusive manner, according to Tononi 2012. These are the following:

Table 2 – Projection spaces of the individual

Projection space	Focal plane
Environment	External reality; physical, chemical and biological processes in the environment, occurring as a consequence of natural laws
Body	Internal reality; physiological processes in the body, occurring as a consequence of natural laws; includes the processes of the autonomic nervous system
Brain	Evaluative focal plane in the attention assessment controller

So far this has been a short summary of how the conscious experience of the individual can be explained, and what are the focal planes, on which informational patterns are basically resolvable. But how can this model be expanded into the social dimension?

This is simply done by involving a further projection space – the society:

Table 3 – The social projection space

Projection space	Focal plane
Society	Social interaction and communication processes; social phenomena, occurring as a consequence of the laws of social interaction, communication and team dynamics. Let us provisionally call this the "social focal plane".

It should be mentioned that the projection space environment is used in a specific way to realise social interaction and communication processes. That is why social interaction and communication always corresponds to interaction with the environment, but with a specific sort of addressee – fellow human beings.

This leads to a combined definition of individual and social consciousness which is very simple and stringent at the same time:

- Individual consciousness is evaluation in the attention assessment controller in relation to the environment and the body.
- Social consciousness is based on this, but applies to multiple people, and additionally involves the social focal plane. So it is individual consciousness from above, but for many people and in combination with relations of the type attention assessment controller – social focal plane – attention assessment controller.

This also leads to the conclusion that the processes that are ongoing in the social projection space, and the patterns that are projected on the social focal plane are the external aspect of social consciousness.

An interesting point in this connection is, that language,

including semantics, phonetics, speech perception and articulation, is interconnected with the associative/cognitive basal ganglia circuit rather than with the circuit for skeletomotor control – see also section "Basal ganglia (BG) and language" in chapter "Basal ganglia (BG) and frontal cortex" in Part 2; see also Chan, Ryan and Bever (2013).

This may have the following consequences:

- Language, communication and social consciousness can be managed independently from and in parallel to skeletomotor control, so that there are few conflicts between activity on the one hand and thinking, planning, communication and social coordination of conscious perceptions and intentions on the other hand. Humans are well-optimised to manage action and communication at the same time in parallel.

- Language is closely bound to declarative memory, associative evaluation, preparation and decision making processes and thus to individual consciousness, so that it can function as direct line between conscious contents in the brains of different people. It is likely that internal linguistic processing, but also information transfer between individuals via articulation and speech perception, works widely without noteworthy confusion with further activity control, as, e.g., walking or working with the hands.

- Internal associative and cognitive processes might be connected with linguistic semantics to a great extent. This means that individual thinking might often occur as a kind of linguistic simulation with downregulated vigour and inhibited motor control. This means also that the threshold to turn into real communication (by talking or writing) is always very low, so that the intellects and feelings of humans in a group are naturally as closely related as ever possible.

But communication and alignment of conscious intentions between individuals and so of social consciousness is always difficult and occurs with limited precision. A broadly precise reflection of informational patterns on the specific human focal planes, in the sense of differentiation, integration and exclusion, according to Tononi 2012, can only be achieved in a provisional manner. There are reasons why conscious information processing and, also, conscious information transfer is always largely imperfect.

Informational patterns are only compatible with one another as long as this is according to the same type of projection space and focal plane. This means that informational patterns of the environment are only compatible with informational patterns of the environment and neuronal excitation patterns are only compatible with neuronal excitation patterns, but that there is no compatibility between environmental and neuronal informational patterns.

This means, further, that compatibility always needs a double transformation – from one projection space to another and vice versa. This means that there is:

- a general incompatibility between patterns for the entities *environment*, *body*, *cerebral cortex* and *social expression* and
- a gradual incompatibility between specific informational patterns that are related to the same entity in a basically compatible manner, but caused by losses during a double transformation.

That pattern transformation between environment and environment via cortical incarnations of the patterns, or between cortex and cortex via social reflection, can occur in a highly precise manner is not self-evident. The usual laws of nature typically take effect in the opposite direction, yet as another form of evolutionary pressure, by always undermining precise forms of reflection. But the attention assessment controller, with its purposeful working mode, is able to withstand the forces of this kind of pressure. As a specific occurrence of the laws of nature, it is the only instance which is able to require that a highly precise projection is enabled for the informational transformation processes. This is one of the most important optimisation tasks that have constantly to be managed by the brain.

But this aspiration is undermined by further limitations, besides the ones that have been mentioned above. Typical types of losses are the following:

- Transmission errors: Erroneous transmission and transformation of codes or signals, e.g. in the sense of a literal misinterpretation of spoken language. Possible errors of addressing, e.g. the wrong person feels addressed.
- Semantic misinterpretations, e.g. due to differences in mental profiles or cultural conventions.
- Contextual losses during activity control or situational preparation. Each specific event is always embedded in a broad context – it is actually the state of the environment around the event or of the entire world, in this moment. The first problem is that only a limited selection of contextual information can be transferred, together with the actual information pattern, which represents the event which is in focus. The second problem is that the context is changing while the information goes through a double transformation.
- Contextual losses during thinking or creative preparation. The contexts of different informational patterns, which are combined by any kind of creative preparation, are always incompatible, to a certain extent, because of distances in the topical, spatial, temporal or realistic sense. An exact match of the contexts is never provided and it is always an important task to be considered to evaluate the impact of the contextual differences. But the success of this pursuit depends on the amount and

sufficiency of the contextual information which is available, along with the actual information pattern, and thus also on the general capability to deal with complexity (see also chapter "The 4DI model" below).

- Focusing to limited topics and details: The chances for high-precision information processing are higher, the more detailed the thinking and the larger the bandwidth of the involved aspects. But the ability to manage important information comprehensively is always more or less limited, dependent on boundary conditions and mental states (more details see in chapter "The 4DI model" and especially in section "The tunnel vision paradox").
- Unconsciousness: There are also always subconscious procedures and information transfers, and large parts of the natural laws are unknown, with the result that extensive parts of the processes are always hidden. They are typically hidden to all the communication partners involved, but there should also be many cases where some aspects are hidden to some participants, and not to others.

A general conclusion should be that limited compatibility and precision, as well as misunderstanding, is the rule for conscious information processing at a social level, and that accurate high-precision communication is, rather, the exception, which, to cap it all, has to be eked out again and again by major optimisation efforts of the involved brain processors.

The 4DI model

A four-dimensional intelligence concept (4DI)

Explanations of the brain lead, among other things, consistently to the question: "What is intelligence?" Intelligence is traditionally associated with accurate information processing, and thus with categories like cognitive capabilities, informational resolution ability and intelligence quotient (IQ). At the same time, intelligence is also associated with other concepts, like emotional intelligence and multiple intelligences, but there is no integrated concept, which tries to combine all the aspects involved in a consistent concept.

In each case, it is typically (traditionally) assumed that there is a "dichotomy between passion and reason" (Descartes) or a contrariness between cognitive capabilities and emotions, and that emotions cause cognitive biases and thus reduce intelligence in the cognitive sense. But there is also an ongoing discussion, which shows that this contrariness book is increasingly perceived as being challenged – see also section "Seesaw models and more advanced approaches" in chapter "Cognitive control and emotions" in Part 2.

However, the SPP model suggests (here) a rather integrated concept, which tries to combine all aspects involved in a consistent manner. The book of a contrariness between emotion an intelligence is turned to the opposite – to the assumption that intelligence is rather based on the capabil-ity to combine both aspects. According to the SPP model, intelligence is seen as referring to the complexity stress field

between four aspects or dimensions, which might be named as follows:
- "informational precision",
- "informational bandwidth",
- "emotivational bandwidth" and
- "team intelligence".

The former three represent an intelligence concept of the individual, the latter makes social intelligence of this.

According to these four dimensions and to the assumption that contradictions and problems may always occur in the complexity stress field between these four dimensions, intelligence is here defined as follows:

Intelligence is the capability to manage complexity in the 4D tension field.

How these four dimensions can be defined and how 4D intelligence (4DI) evolves in the stress field between them, will now be outlined.

(1) Informational precision
(2) Informational bandwidth

This part of the concept comprises all kinds of information processing capabilities of the brain. It refers to the abilities to interpret, combine and store complex information in a successful manner, and to generate reasonable behaviours and knowledge from this. This is the kind of intelligence that is focused by classic intelligence tests. It correlates with the interconnection complexity of the cerebral cortex and depends on this.

Detached from the discourse on this matter, the boundary conditions for this aspect of human intelligence can be defined as follows:

- These two informational dimensions focus on all kinds of perception and information processing capabilities of the brain.
- The information does not have any meaning or consequences for the states and conditions of being.
- It includes neither motivational aspects nor emotions (besides, possibly, ambitions for playful testing; note: motivational and emotional aspects are rather assigned to dimension (3), as described below).
- It applies especially to game-like situations and to performances which can be investigated by tests (e.g. intelligence tests).
- Cognitive capabilities are included, but only in terms of information processing capabilities. Speech and communication are included, but only in terms of information exchange and the acquisition of knowledge at a social level. Purposeful and empathic aspects, which are also an inevitable part of cognition and communication, will be regarded as excluded in this case (note: this is rather assigned to dimension (3), as described below).
- Multiple aspects, complex interdependencies and cross-modal associations are included, referring to the multifarious requirements that have to be managed in real life, to interdisciplinary approaches and to comprehensive knowledge, but only in the informational or intellectual sense by excluding drastic experiences, traumas and emotivational aspects. This point refers specifically to dimension (2), informational bandwidth.

Dimension (1), informational precision, stands for: precision, sensitivity, IQ, accuracy, scientific correct results.

Dimension (2), informational bandwidth, stands for:

knowledge, amateurish and professional ambitions in any practical or theoretical fields, interdisciplinarity.

The two aspects or dimensions – (1) informational precision and (2) informational bandwidth – involve already an objective type of contrariness and area of conflict. This results from the general inability to resolve all aspects of the world with infinite precision or to acquire total knowledge on the basis of limited resources. Gnosis is always a compromise with limitations. It can always only comprise a special aspect of the world and this aspect can only be resolved with limited precision or resolution.

This is reasonable when taking a look to Google Maps, for instance. One can either see the globe as long shot, but only with limited resolution or one can zoom into a concrete region, seeing all the details with rivers, villages and towns, lakes, mountains and oceanic coasts. But it is never possible to combine this on a screen with a limited number of pixels.

Of course, a human can zoom into different areas of the world and then again out of it, returning to the long shot perspective. And then he can imagine that there are similar structures all over the world, consisting of rivers, villages and towns, lakes, mountains and oceanic coasts of different kinds, and he can make conclusions from this, based on abstractions. But the results of this combination might not be totally correct, because abstraction means fading out variations and exceptions, which might be present in a part of the world which has not been examined in detail. And the human cannot examine all areas in the world in detail, due to limited time, and even if he could, his brain would not be able to capture all the details.

Of course, the Google computers manage the totality of detailed information, which has been collected for Google Maps, but they are less well-equipped with cognition abilities than humans (the former are actually not equipped with any rudimentary piece of cognition), and the Google maps are not as detailed as any other maps, like for instance some OSM maps. And even if this is overcome at any time in the future by artificial intelligence and big data solutions – this would not solve the contradiction, because this is also always based on limited computation capacities and on the limitation that only types of data can be involved that have explicitly been collected for this purpose. However, there could be a potential to challenge humans by computers in these two dimensions.

It should be noted that contemporary concepts of cognition and intelligence (in the informational sense) apply typically more to the informational precision dimension (1) in one-sided manner, rather than to the combination of both dimension (1) and (2). Any tests, as e.g. IQ tests, focus usually on small problems and tasks, and it is checked, to which extent they can be solved accurately. This means distinctive focusing on small isolated aspects of the world, and thus extreme underrepresentation of dimension (2) *informational bandwidth*.

The result is that our understanding of intelligence is widely aligned to accurate cognition abilities of very small aspects of the world, while real life is different – it challenges always by a broad variation of aspects and by the necessity to draw combined conclusions in a more or less provisional manner.

This area of conflict gets even more confusing when involving the next two dimensions.

(3) Emotivational bandwidth

The next dimension arises when considering that life follows always purposes that are involved via homeostasis, sensation of pain, emotions, motivations and any kinds of needs or rewards. This is involved in the brain via emotivational signals and excitation patterns, which are the set values in the evaluation process of the attention assessment controller and which are included as assessment values in stored memories, so that they can be involved as value-added solution patterns into the evaluation process at any time later on.

The consequence of this dimension is that needs have to be managed, and deficiencies and crises have to be combated. Information processing can no longer follow any kind of whim or indulge in playful testing or pure aesthetic enjoyment. Rather it has to focus on awkward truths, introducing conflicts and stress fields arising between senses of delight and elation, on the one hand, and fear and deprivation, on the other hand.

Dimension (3), emotivational bandwidth, stands for: passion, empathy, steadfastness, pertinacity, resistibility, vigour, mental power (also physical power), emphasis.

This dimension has a contradictory relationship to both dimension (1) and (2). High informational precision and good resolution ability (1) means among other things that the potential to solve problems and difficulties quickly and efficiently is extraordinary high. But it is clear that there are always lots of remaining scenarios, where this approach does not work and where difficult situations turn into longer lasting events. And this is exactly the type of scenarios where consciousness arises and real intelligence is needed to overcome the issues. If it went already this way, high resolution ability (1) is now rather more disadvantageous than beneficial. This is true because hardship, once occurred, is felt the more painful the higher the resolution ability, and thus the higher the sensitivity of the neural apparatus. The good abilities and rich solution pattern equipment may turn again into an advantage later on, when a solution is found with its behalf, but the stress fields which have to be suffered in between, are the higher, the more complex the neural networks and the larger the neural sensitivity and resolution abilities.

The same is true for the informational bandwidth dimension (2). Large amounts of information, extensive knowledge and multifarious ambitions are a good basis for quick and efficient solutions as well as for the ability to find good final solutions due to longer lasting difficulties. But in the meantime, when a situation turns into a longer lasting problem and when the solution is stuck for any reason, they are more an impairment than an advantage. This is true, because the more information and knowledge is available in the positive sense, the larger also the information and knowledge about risks and dangers which may threaten in fail cases; and the latter painful aspect has more weight, when there is no way out for a certain time period.

Particularly in cases of longer lasting problems, which cannot be solved immediately, it is the ability to plan and act with high grades of consciousness, control and intelligence, which is the decisive resource to overcome the difficult situations. This is the type of situation where intelligence belongs to as crucial resource. Even if an individual is equipped with rich and well-suited associative-emotivational and sensorimotor-emotivational patterns, and if he or she is able to meet all demands of real life with bravery for a certain time period, it is guaranteed, due to the ever-changing environment, that new situations will come where new strategies must be learned. And this is exactly the time when real intelligence has to prove itself.

That's why intelligence, at least of the 4DI type, means the ability to suffer emotivational stress fields and to nevertheless bring good results in cognitive disciplines. It is the ability to deal with extraordinary high complexity in the area of conflict between informational resolution ability/precision, informational bandwidth and emotivational bandwidth. But it should be clear that difficult times or crises are characterised by extraordinary high emotivational complexity or bandwidth, respectively, with the consequence that this must be compensated by downgrading the other two dimensions – informational precision and bandwidth –, and thus the degree of rationality.

The demands to manage complexity become much more difficult with the next dimension.

(4) Team intelligence

The fourth dimension is also achieved by humans, but is beyond the potential of a single individual. This is the ability to get the results of efforts in relation to dimensions 1-3 implemented in social processes and achievements.

Team intelligence means that control functions that are managed in the brain of an individual with the behalf of his attention assessment controller on control levels three to five, are on social level synchronized between multiple people, resulting in coordinated efforts for shared goals. This is linked with concepts like language, communication and social consciousness, which have been mentioned in the sections above.

The concept "team intelligence" has the following impacts among other things:

- **Shared problem complexity**: Cooperating in a group or in an exchange relationship means that a joint issue spectrum must be managed; each participant involves additional problems with individual shades, what results in a group-related problem complexity stress field in the 3D space, which is much larger than this of a single individual; term "3D space" refers to dimensions (1), (2) and (3) from above. The combined 3D tension field of a group of people might also be called 4D tension field or 4D stress field.

Example: Two neighbours decide to cooperate in caring their pets, a cat and a dog; this brings additional efforts for managing different kinds of food, coping with tensions between the two species and animal individuals and for organising cooperation.

- **Shared potentials and mutual support**: The participants can join their resources, skills and knowledge, so that the potentials of the team are much greater than that of one person alone. And they can support each other.

  Example: One of the two pets is ill (the cat); the two neighbours can join their knowledge in finding suitable therapies and measures; the dog owner may carry the two pets to a veterinarian consultancy; the cat owner enjoys this support while he has to go to work.

- **Roles and responsibilities**: Individuals have often to take the responsibility for the whole team in certain aspects or for certain time periods; this means that a single person has partially to cope with amounts of 3D complexity of the team, thus of 4D complexity, which are much more demanding than his personal affairs.

  Example: The dog owner has to intermediate between the acute pain of the cat, the therapy proposals of the veterinary and the potential intentions of the cat owner, while he has also to take care of his own dog at the same time.

- **Social structuring of life**: Interactions of multiple individuals may occur in highly dynamic manner (small talk, payment at a cash register) or with smaller or greater duration (consultation, family celebration, cultural event, any kind of teamwork, a labour contract or any other kind of contract). The team context ranges from very small (two friends, a couple) via manageable sizes (family, work team) to extraordinary large or global (corporation, institution, state, the global economic system). However, the mutual interactions and interdependencies are nowadays extremely rich provided, so that life takes widely place in well-structured manner, and each of us has to deal with 4D complexity forces that are stretched into multifarious directions.

Example: Real life.

- **Artificial person:** A group or team, in the sense of an artificial person, can again be involved into interrelationships like a natural person. In this case, the common 4D tension field and the 4D intelligence of the artificial person might enfold its power like the 3D tension field and 3D intelligence of a natural person.

  Note: As already stated above (in relation to "shared problem complexity"), a 4D tension field is nothing else than a 3D tension field, but with the specialty that it is a set union or combination of the 3D tension fields of multiple people.

  Example: A man has a contract with a big insurance company; he can make great benefit out of the combined expertise of the insurance company; this works well until the day he gets in trouble with them – then it is difficult for him to face their power in a legal dispute; this gets better when he is able to join a bellwether trial which is supported by the power of a large number of complainers.

The result is that life is more characterised by occurrences of team intelligence and 4D complexity than by any other phenomenon, or that all more concrete social phenomena can, on an abstract level, be interpreted as processes, which are ongoing in the 4D space.

The assumption that there is "a 'dichotomy between passion and reason' (Descartes) or a contrariness between cognitive abilities and emotions, and that emotions cause cognitive biases and thus reduce intelligence in the cognitive sense" (see the second paragraph of this section), which has been supported by traditional seesaw models, contains still some truth, also with the 4D intelligence model. But a crucial point is that the 4DI model is based on the book that cognitive performances get their sense only via purposes that are incorporated from the emotional-motivational system. There would be no cognition without this component, be-cause cognition without goal is only informational chaos. That's why the 4DI model sees cognition and information processing as always, at the same time, driven and impaired by emotivation and passion, what might not be asserted by other models, particularly traditional models, with such defi-niteness.

With the 4DI model, the greatest merit of the brain lays in its ability to combine information processing with purpose and self-regulation. This means that there is no information processing which is completely free from impairment by emotivation – only the degree of the impairment scales from huge to very small. But this means also that vigour is always injected into the brain processes to an extent which can widely be adapted to the demands in each situation. Exactly this ability should be the greatest accomplishment of human intelligence, not cognition alone.

Finally, the features of human intelligence, as promoted by the SPP-4DI model, can be summarised as follows:
- Self-regulation support for the existence of human individual and species.
- Activity control, cognition and reason on the one hand, vigour, mental power, passion on the other hand, and the ability to adapt between both virtues over a wide range.

- Fast unconscious suitability probability processing on the one hand and slow conscious suitability probability processing on the other hand, and the ability to adapt between both virtues over a wide range and to manage smooth handover and reciprocal interaction between both virtues (see also section "The two types of consciousness", Singer 2004, Kahneman 2012 and section "Kahneman, Daniel: Thinking, Fast and Slow" in chapter "Psychology" in Part 2).
- Development and usage of effective and smooth behaviours and innovative problem solutions on the one hand, management of major deficiencies, in terms of internal resources (neural patterns) and external (material) resources, on the other hand, and the ability to adapt between both demands over a wide range.
- Projection of all important aspects of the universe into an internal associative-emotivational pattern library, leading to an internal view of the world with a certain grade of coherence. Constant development and optimisation of this world view and alignment of problem solutions with this world view (see also section "The need for coherence" below).
- Integration of all informational aspects of the self-regulation process and of memorised patterns into a coherent value system, which allows value-guided decision making and balancing between all aspects involved. The result is a universal assessment and evaluation system with potential inclusion of all aspects of the world.
- The ability to manage life via the detour over artificial needs, aesthetic evaluation and cultural accomplishments rather than only by a constant combat against deficits (see also the next section below).
- Development, optimisation and strategic coordination of all these features on social level, making social con-

sciousness and team intelligence of it.

The concepts "4D tension field" and "4D intelligence" might provide good opportunities to investigate the implications of all these components of intelligence on an abstract level in integrative manner.

The so called "artificial intelligence" (AI), as it is projected nowadays, lacks nearly all features from above; it might only refer to some specialised kinds of data collection, data transformation, pattern recognition, control and cognition, but which are implemented separately from one another and only combined by pieces of conventional software.

The question of whether a system can be attributed as intelligent depends on the fact that it can develop universal integrability and extensive adaptability. The human brain is such a system. The so called "artificial intelligence" can only exist by depending on the universal integrability and extensive adaptability of humans.

Dynamics of the need hierarchy

Note: This headline has been adopted from Maslow 1987, 17; more details see further down.

That needs are resulting from emotivational signals and patterns and that needs play an important role in the self-regulation processes, has been mentioned in several sections above. The concept "need" is just another term for all kinds of patterns that are emitted by the emotional-motivational system in the brain and which have mostly been called "emotivation" in the sections above. Thus, a need is just an emotivational pattern which takes effect as (potential) goal in the decision making processes of the attention assessment controller (AAC).

The intention is now to turn into a more detailed view

on "types of emotional-motivational neural patterns" or "needs", respectively, at which the latter term is significantly better suited for this aspiration than the former; this is the case for two reasons: term "needs" is smarter and it is a good connection between neurosciences and psychology.

Needs have already been categorised into basic needs and higher or artificial needs in the sections above. The former are emitted by the autonomic interface and by pain sensation, and their regulation occurs typically in the sense of remedying deficits. The latter are emitted and maintained by the mesolimbic dopamine system (reward system), and their regulation occurs typically in the sense of receiving a reward, achieving a goal or making any good experience. There might be further mechanisms, which have been named "emotivational dependency chains" and "emotivational stimulus modulation" (see section "Higher needs"), which might represent connections between both major categories. Emotions and feelings in the range from anxiety and fear to delight and elation are – facilitated by the amygdala – involved in either case.

The need hierarchy specification in figure 2 and table 4 zooms into a more detailed categorisation of needs. It is based on Maslow 1987 and on several popular variants of the hierarchy of needs, and it is here provided in a modified version. This categorisation, as well as any other attempt to bring this matter into a schematic order, can only lead to provisional results. But it is provided here because these more concrete examples might be a good basis for a better understanding of the topic and also for a more concrete discussion of some dynamic aspects.

The need levels 1 to 5 in table 4 may correspond with the major need categories, thus basic needs and higher/artificial/virtual needs, as follows:

- Level 1 should have to be attributed to the major cat-

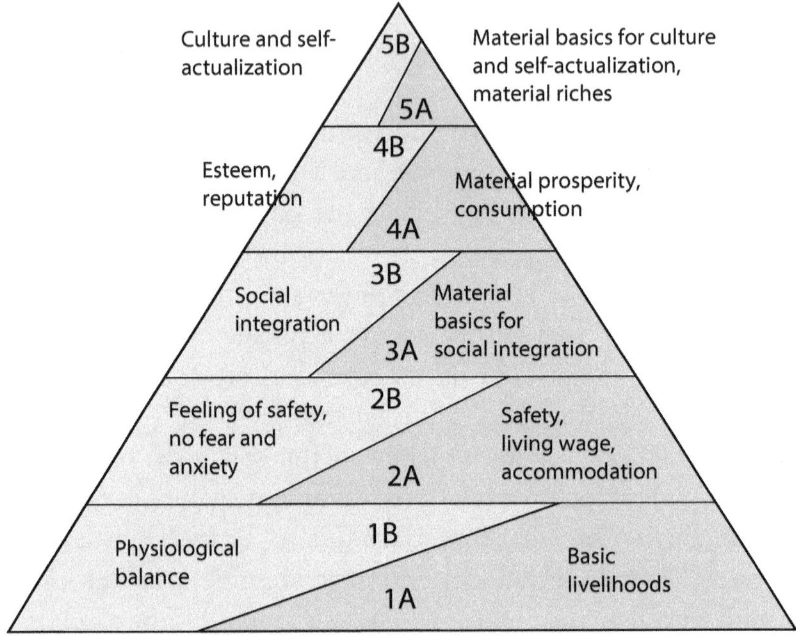

Figure 2 – Pyramid of needs

egory *basic needs*. Homeostasis and pain refer always to urgent necessities, and have typically to be solved immediately in the sense of remedying deficits.

- Level 2 refers already to living conditions and citizenship and so to the labour market, the residential market, the enforcement of law and order and cultural conventions. This involves a certain room to manoeuvre, referring to the development of *artificial needs* in several directions, but the resulting concrete needs should still be very close to basic needs. It is a kind of basic artificial needs.

- Level 3 involves the social aspect of life. This can be seen as strong interconnection with all need categories, which are involved on social level via communication, social consciousness and social cooperation. Belonging-

Table 4 – Hierarchy of needs with aesthetic and material aspects

B – Aesthetic aspect	**A** – Material aspect
5 – Culture/self-actualization/growth	
Need to constantly develop one's capabilities and talents, need for knowledge and understanding, need to be able to go one's own way in the social environment; need to make a positive contribution to the welfare of fellow human beings, the development of human society, environmental protection; religious and spiritual needs; need for culture and, perhaps, for enlightenment	Sufficient material basics for culture and self-actualization, material riches
4 – Individual needs/esteem need	
Esteem, reputation, prestige, recognition (awards, praise), influence, personal and professional achievement, mental and physical strength, stable self-esteem	Material prosperity; financial means, tangible assets, fashion, consumption – for personal use and to symbolise status and prestige
3 – Belongingness needs/social needs	
Family, friends, partnership, love, intimacy, communication, social integration and cohesion, belonging to a community	Sufficient material basics to be able to maintain family ties, friendship, partnership and all other kinds of social relationships and social integration

B – Aesthetic aspect	**A** – Material aspect
2 – Safety needs	
Feeling of safety and physical integrity; no fear and anxiety; legal certainty is guaranteed; need for structure	Safety, stability, law and order; protection against danger; fixed minimum living wage; safer accommodation
1 – Physiological needs/homeostasis	
Hunger, thirst; need to breathe, sleep, sexuality, exercise and activity; temperature balance, homeostatic balance; no pain	Food, water, fresh air; sufficient possibilities for sleeping, sex, exercise and activity; beneficial surrounding temperature, sufficient preconditions for complying with homeostatic regulation; physical integrity of the body

ness needs are, on their own, *artificial needs*, which may mostly also refer to artificial needs of fellow human beings at levels 2 to 5. But they can also involve *basic needs* which are suffered by people who need help.

- Levels 4 and 5 are clearly composed of nothing else than artificial needs or growth needs, respectively.

These five levels or the levels of any other hierarchical need model are not strictly divided. There should rather be lots of vertical relationships which might be caused by mechanisms like "emotivational dependency chains" and "emotivational stimulus modulation" or by other impacts. An appropriate phenomenon should be, that each need type from an original level gets corresponding representations on each level above.

Examples for this phenomenon might be the following:

- The need for food and water is originally assigned to level 1. But it has incarnations on each level above. It might be cultivated as food hoarding on level 2 (safety), family meal on level 3 (social), working dinner on level 4 (esteem, reputation) or gourmet experience on level 5. Or some of these demands are sought in combination.
- The need for safety and stability is originally assigned to level 2. But it may have caused the need for stable social relationships – as e.g. long term friendship, family ties etc. – on need level 3, the need for a professional career and material property on need level 4 and the need for a coherent internal world view in order to face the challenges of life on need level 5.
- Sexuality is originally assigned to level 1. But it could possibly also be assigned with the following phenomena at higher levels: marriage in relation to level 2 (safety) and 3 (social), beauty in relation to level 4 (esteem, reputation) – either own beauty or beauty of the spouse or both –, sexual pleasure and erotic adventure in relation to level 5 (aesthetics, self-actualisation), sexual morality of religions in relation to level 5 (religion, spirituality) – as e.g. no sex before marriage, celibacy.
- Social relationships and needs (level 3) are always an important basis and motivation for the cultivation of higher needs (level 4 with esteem and reputation, level 5 with culture and self-actualisation etc.).
- Esteem, reputation, influence (level 4) are closely connected with cultural ambitions (level 5).
- That there are strong dependencies on the material side of the hierarchy – from basic resources (level 1) over safety (level 2) and sufficient amounts of money and property (levels 3 and 4) to wealth (level 5), is a matter of course anyway.

This type of interrelationships between need levels might also be a mechanism which enriches higher need levels with emotivational phenomena like passion, empathy and power.

Now let's turn to the "Dynamics of the Need Hierarchy", as described in Maslow 1987 at pages 15ff. According to Maslow, "the basic human needs are organized into a hierarchy of relative prepotency" (17). This means that basic needs, if unsatisfied, dominate the organism. "For the human who is extremely and dangerously hungry, no other interests exist but food. He or she dreams food, remembers food, thinks about food, emotes only about food, perceives only food, and wants only food." (16f.) But this turns into another state of the organism when hunger can be satisfied. "It is quite true that humans live by bread alone – when there is no bread. But what happens to their desires when there is plenty of bread and when their bellies are chronically filled? [...] *At once other (and higher) needs emerge* and these, rather than physiological hungers, dominate the organism. And when these in turn are satisfied, again new (and still higher) needs emerge, and so on." (17) "The physiological needs, along with their partial goals, when chronically gratified cease to exist as active determinants or organizers of behaviour. They now exist only in a potential fashion in the sense that they may emerge again to dominate the organism if they are thwarted." (18)

The consequence is that there is a steady interplay between lower needs and higher needs over all levels, at which lower needs tend to be prepotent, if active, and higher needs develop only on the basis of the satisfaction of lower needs. This is a principle over all need levels. How concrete needs are specified, categorised and ordered into a hierarchy is a rather secondary topic. The more important message is that needs build on each other in a hierarchical interdependency and that lower needs tend to occupy the ambitions of the organism, if active, while higher needs evolve and develop on more subtle levels in consequence of the fact that lower needs are managed and satisfied successfully.

This fits very well to the fact that self-regulation occurs in the brain on the basis of emotional-motivational signals and patterns, and to the theses that higher/artificial/virtual/growth needs develop on the basis of the reward system, *emotivational dependency chains* and *emotivational stimulus modulation*. The crucial point here is that the raw material for higher needs are pattern combinations that are constantly produced by the processes of the attention assessment controller (AAC), and which include always informational and emotivational components. Thus higher needs, which evolve from these pattern combinations, are always composed of fragments of basic-need patterns or emotivational patterns, respectively. This results in emotivational patterns which are the more complex and subtle the higher the need level. In this sense, higher needs are only weak and ambiguous incarnations of lower needs. Reversely it can be concluded that lower needs, which are the more clearly straightened originals, are, if occurring again, always *prepotent* in relation to higher needs.

This is easily understandable, when thinking again of the example of a "chronically and extremely hungry person" from Maslow (1987): "Freedom, love, community feeling, respect, philosophy, may all be waved aside as fripperies that are useless, since they fail to fill the stomach. Such a person may fairly be said to live by bread alone." The higher needs as mentioned above, as well as all kinds of cultural ambitions, might not least be developed to make hunger – or other lower needs, as pain and safety – to rare events, but it is clear that the actual deficiencies have regularly priority if occurring again despite the precautionary efforts. This relationship occurs at each higher and more subtle need level again – scientific truth, e.g., which can be attributed to need level 5, is nowadays an important means for the ambition to manage the issues of life in well organised manner, but it is pushed into the background if social or financial trouble has to be managed (at levels 2 to 4) or even if reputation is concerned (level 4).

The principle of the relative prepotency of lower needs in relation to higher needs corresponds also very well with the four-dimensional intelligence concept. This is outlined in section "Dynamics in the 3D tension field" below. But we will firstly discuss two further phenomena of the need system – "social emotivational dependency chains" and "the need for coherence", and we will clarify in what way concept "growth needs" is appropriate.

Social emotivational dependency chains

The concept "social emotivational dependency chains" is the social extension of "emotivational dependency chains" as described in section "Higher needs". It bases on the following mechanisms:

- The extraordinary ability of the human to produce controlled behaviours and attitudes via the attention assessment controller might preferably be developed for regardfulness towards the social environment and fellow human beings rather than towards the natural environment (more details see under "Considerations on the social reference of impulse control" in chapter "Basal ganglia (BG) and frontal cortex" in Part 2). It can at least be assumed that social forces tend to have priority over natural forces, except for the basic need levels 1 and 2, which occur only exceptionally in a well-organised society. This high degree of social determination is also the reason why the social need level is the lowest at the transition from deficiency needs to growth needs.

- The extraordinary communication abilities of humans and the capability to develop social consciousness (more details see in section "Individual and social consciousness" above). This includes namely also the emotivational aspects of life, so that emotions and feelings of different people are connected and made interdependent on each other.
- And even the *emotivational dependency chains* (inside of individuals) as already mentioned above.

These mechanisms lead to the peculiarity of the human society to interrelate individual emotivational dependency chains in multifarious manner to one another, and to build a network of social emotivational dependency chains by this way. It can be assumed that life in a well-developed society is determined by these dependency chains to a great extent, or that this is widely the basic mechanism of how needs evolve in individuals that are integrated into the society.

Thus, social needs and *social emotivational dependency chains* can be seen as the basic need mechanisms for people who are well integrated in a well-developed cultural society. Real basic needs, so at level 1 and 2, occur only for exceptional cases, so for individuals who get sick or who are (partially) excluded from normal life for any reason, or for the case that the social or economic conditions get into difficulties.

However, the society as a whole will always have to manage also the need levels 1 and 2, what can only function if representatives suffer this type of needs and share the corresponding feelings. This occurs, for example, by the following ways: sickness and exclusion (as mentioned above), social, economic or political distortions (challenging the attribute "well-developed"), professional or voluntary confrontation with basic needs (hard jobs, sports), adolescence (in terms of structural development), confrontation with greater environmental changes and challenges, societies that are less well-developed or that get under pressure due to changed boundary conditions, serious crises, catastrophes and wars.

The humanity would lose its purposefulness and orientation if these mechanisms are overcome, but this would certainly result in the next crisis, so that pronounced emotivational impacts are again injected into social mentality.

The need for coherence

A special need should be mentioned, which could be entitled by different names. We decide to call it "need for coherence".

This need can be seen as a general demand, which covers all other needs and particularly the artificial or growth needs in an abstract manner. It arises from the experience of basic needs – particularly the needs for safety, integrity of the body and homeostasis – in combination with the experience that precaution can be managed in a good way via the detour over artificial needs, so that further basic experiences can be avoided to a great extent. This can only be successful if the artificial needs are developed and cultivated in a systematic way. Thus, it is consistent that the need for safety and stability from level 2 turns, among other things, into a distinctive representation on level 5, which requires that the complete system of artificial needs and its actualisation follows any kind of order, which is associated with guaranteed future. This is more or less inevitable, only the severity and specificity might vary over great ranges between individuals.

Some more concrete incarnations of this *coherence need* can be described as follows:

• This might be the actual essence behind all artificial

needs that are assigned to level 5 in table 4, as e.g. culture, self-actualisation, need to make a positive contribution to the welfare of fellow human beings etc.

- Need for aesthetics, aesthetics principle or principle of differential aesthetics: The demand to experience aesthetic elations as experiences which represent the greatest distance from deprivations.

- Library need: The demand to maintain the library of associative-emotivational patterns and sensorimotor-emotivational patterns in a way, so that it may always provide suitable solutions for any tasks or problems which can occur in present and future. How this appears depends widely on age, boundary conditions and mental characteristics. This might be lived out as play instinct and inquisitiveness during childhood and adolescence – when the neural pattern library is like a dry sponge, which is greedy for absorb everything –, as need for professional or amateurish competencies and for a coherent world view during adulthood or as need for keep a remaining level of control, as good as still possible, during the process of deceasing.

- Believe, religion and irrationality: Humans must always deal with certain gaps in their competency matrix and world view. It is actually a fact that the plenty of gaps is rather huge while the pieces of expertise, skills and knowledge are very small, compared to what is possible in the world. This is a massive assault against the need for coherence, making coherence impossible on a realistic way. But the strategy of the human brain is to treat these gaps by tailored harmonisation filters, which make them as small and harmless as ever tolerable. Humans are great masters in making such types of harmonisation filters, and the results are all kinds of believes, religions, irrational attitudes and biases. In this sense, the ability to endure the challenges of life and keep optimistic and act with verve at the same time depends more on the ability to ignore real risks and contain threats by obscure believes and pieces of irrationality, than on knowledge, skills and competencies in the positive sense. This fits also to the finding or book that *emotivational bandwidth* and thus irrationality plays a crucial role in human intel-ligence (see the 4DI concept).

The coherence need, as it is described here, can be seen as an abstract general goal or as the neural fulfilment principle, which is the actual truth behind each more concrete demands and efforts of humans. It is the ultimate driver of culture and affluence.

See also Singer 2002: Under the headline "Attachment problem" ("Bindungsproblem"), he discusses of how the "self" ("ich") might be constituted, and he postulates that **coherent** interpretation of the multifarious sensational signals must occur anywhere in the brain. This refers more to consciousness and the search for the "self", but has also a link to the *need for coherence* as it is proposed here.

Artificial needs versus growth needs

Artificial needs can be attributed to the human or Homo sapiens as a basic feature from the beginning. This might also apply to animals, as e.g. mammals, even if with limited extent. All species with a reward system in their brain should be able to develop artificial or virtual needs, so needs that are not genetically determined and that are not necessarily bound to body conditions. Pavlovian responses should already be simple pre-stages of artificial needs.

But in what way can there be the talk of growth needs? Artificial needs might have turned into growth needs during human history. This might have started as inquisitiveness or collection passion of some people over many thousand years. But it encompassed the whole society at the latest since the 16th century. Since that time, the passion towards growth became the most significant social movement of our era. This occurred in the form of the scientific technological revolution and economic growth (see also Harari 2014, chapter 14 – "The discovery of ignorance"). That's why it seems nowadays obvious and hardly controvertible that growth needs are an important component in the human need system, which has large influence on individual and society, at least as a modern tendency. But this is not an inherent ingredient of human nature, like artificial needs.

It should be mentioned that mental growth must not necessarily be bound to economic growth or increasing consumption of resources. Real mental growth might rather correlate with increasing independence from sensitive goods, so with de-growth in the material sense.

Dynamics in the 3D tension field

The four-dimensional intelligence concept, as introduced above, bases primarily on the 3D tension field between informational precision (IP), informational bandwidth (IB) and emotivational bandwidth (EB). These three dimensions stand, at the same time, in a complementary and contradictory relationship to each other. Intelligence of the human kind can only occur, if one is able to last the tension field between the three dimensions to a certain extent, so that a large range of aspects, high informational precision and mental vigour can enrich one another.

But the fact that the three factors or dimensions oppose each other at the same time, means that the ability of humans, to deal with this complexity and to exploit the provided opportunities for their benefit, is also always limited. It is possible to deal with many details (IP) or aspects (IB) or with profound emotivational stress (EB), but it might never be possible to exploit all three dimensions at the same time to an extraordinary high degree. This is prevented by the fact, that the neural resources are always limited. This means more concrete, that the product IP * IB is restricted by memory capacity, and that the product information * emotivation, and thus memory * emotivation, is restricted by the limited ability to suffer pain.

A conclusion can be that the "mental capacity" of a human (or animal) is defined as follows:

$$MC = IP * IB * EB = constant \text{ (short term)}$$

Further conclusions might be the following:
- It is basically possible to develop ones mental capacity (MC), but this occurs, if at all, only long term. Examples can be: long-term development in brain memory capacity and cognitive capabilities over generations, changes in frustration tolerance due to experiences and personal development.
- Short term adaptation is only possible in the sense, that one factor is regulated up or down more or less quickly and extremely, but only on the cost of compensation in the remaining two dimensions. This is outlined in more detail further down.

Conclusions in relation to the fourth dimension, team intelligence:
- The formula and rules from above refer also to the 3D complexity tension field of teams and to social consciousness. This means that the mental capacity of individuals refers inclusively to the team contexts in which

they are involved, and that teams have also a mental capacity, which takes effect in relation to the combined 3D tension fields of all people involved.

- It can further be concluded, that the mental capacity, and so the 4D intelligence, of individuals, is widely determined by their social relationships. This means that it depends significantly on the social relationships which occur by chance and on social dynamics, which develop due to natural (social) laws, but also on the ability to manage the team dimension in a fortunate or target oriented manner.

If one accepts the formula and rules from above, and the fact that mental capacity is limited, the question is now, what does this mean: *dynamics in the 3D tension field* or *short term adaptation*? How are the three factors or dimensions interconnected to each other, which scenarios may occur in the 3D space and what are the possible consequences?

This topic can be examined by starting from typical reactions to pain, frustration or dangerous situations. If an event of such a kind occurs, the control processes in the brain are always intruded by additional emotivational signals, what involves two components: (emotivational) information about the specific deficiency or threat, and a general increase in the emotivational load, which has to be dealt with. The inevitable consequence of this is, that the negative event occupies the brain processes to a certain extent and brings them into the service of the prevention of disadvantageous consequences.

Applied to the MC formula from above, this means that an increase in the emotivational component EB must result in a decrease in one of the informational components, so either in bandwidth (IB) or precision (IP) or both. In the most cases it is likely that this occurs primarily in the IB dimension. Examples are: driving a car very fast forces narrow focusing on the road in front; pain or stress of any type directs attention to the cause and pushes other aspects of life into the background until relief is achieved; a family conflict impairs other ambitions, as e.g. profession or development of capabilities and talents; depression, in the sense of mental capitulation, occurs as an ultimate consequence of constant excessive demands on the mental capacity (MC).

A typical case of focusing in the IB dimension is provided by the usual stress reactions, which include, among other things, dedicated attention on the cause (or the combatant) and states of alert in the nervous system and in the physiological systems, comprising activation of the sympathetic branch of the autonomic nervous system, hormone release (adrenal, cortisol), increased heart rate, blood pressure and blood flow etc. This is can also be accompanied by intensified perception of the incidents, which are now in the focus (increased informational precision/IP). But the decrease in the informational bandwidth dimension should be much more significant (the world around is more or less completely ignored), so that the rise in the emotivational dimension (EB) is compensated by focusing in the informational bandwidth dimension (IB), and the mental capacity limitation (MC) is not exceeded in the end.

Although if focusing in the IB dimension is a typical kind of reaction on rising emotivational load, adaptation can also occur in the resolution/precision dimension (IP). But it is likely that this is a secondary strategy, which occurs rather more long term, due to natural mechanisms. An example of a short-term adaptation in the IP dimension might be the reaction to extreme stress or psychological shocks, which may primarily refer to adaptation in the IB dimension, as described above, but which can also result in lowered pain sensation or unconsciousness in the end, what refers to the IP dimension. Some examples for long-term compensation in

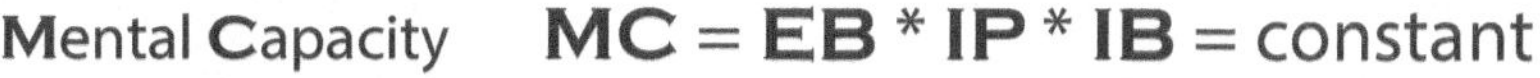

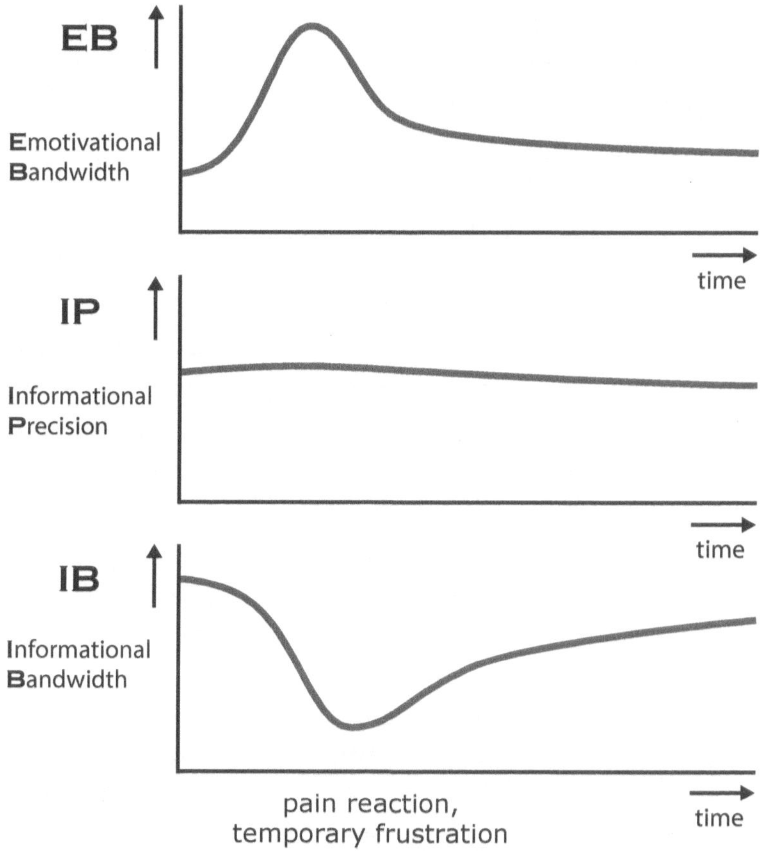

Figure 3 – Primary compensation mechanism in the 3D space

the IP dimension might be the following: decreased cognitive capacity as long-term consequence of depression; unfortunate development of cognitive and intellectual capacities due to childhood in a family climate, which is characterised by larger levels of distress. These types of adaptations seem deplorable, but it should be clear that they are absolutely appropriate under these circumstances.

The following examples may refer to affected informational capacities in general, involving both dimension IB and IP: flight into drugs or dissociative psychological phenomena due to critical levels of distress.

Summarising, one can conclude that new difficulties occur typically as rise of pain, frustration or stress in the EB dimension, and that the brain adapts short term usually in the IB dimension, so by fading out information which is of lesser importance in this situation – see figure 3. This leads to the effect that the efforts of the whole system – so of brain, central nervous system, autonomic nervous system, neuroendocrine system and the physiological processes of the body – are concentrated on the remediation of the cause, what will hopefully result in recovery of the original balance in the end. This can be seen as the most crucial and primary prototype of the compensation mechanisms in the 3D space. However, reality should be full of variants of such kind of adaptation scenarios.

This prototypic compensation mechanism is also consistent to the "Dynamics of the need hierarchy" as described in the appropriate section above. Assuming that an individual is in a balanced state in any moment – so indulging in activity on a higher need level (self-actualisation or maintaining reputation) –, an event at a lower need level – as e.g. social stress or a threat for safety or health – means always that the lower need, which has been raised now, prevails for a while and that the higher ambitions are suppressed. The organism is only able to return to the better balanced state and to the higher-level ambitions, when the stress or the threat could be managed to a sufficient extent and when the lower need could be brought to being vanished again by this way (see figure 3). This kind of dynamics in the need system meets exactly the compensation mechanism according to the MC formula. Higher need levels refer always to informational contexts which are the more complex the higher the need level and the more sophisticated the ambition, and lower needs are always the more unequivocal and urging the closer the demand to the basic fundamentals of life. That's why the statement that a "lower need prevails for a while" (need system) is absolutely consistent to the statement that a "rise in the emotivational dimension (EB) is compensated by focusing in the informational bandwidth dimension (IB)" (4DI/MC).

The result is that the MC formula and the book that the mental capacity (MC) is short-term constant is nothing else than another possibility to describe what is already known as being determined by the dynamics of the need hierarchy.

So far to the primary short-term adaptation mechanism. That adaptation can mean much more, so also adaptation in the IP dimension, long-term change of the balances between the three components or development of the mental capacity (MC), is clear so far and has partially already been mentioned above. This is a field for further investigation, which may promise interesting insights into psychological phenomena.

However, let's now continue with the question: How do the 3D tension field and the mental capacity limitation take effect under circumstances that are widely characterised by artificial or growth needs and beneficial developments?

3D tensions in the affluent society

So far about stress reactions due to rises in emotivational load. But it is time to mention also the other side of this class of phenomena – the boundary conditions for the development of excellent cognitive capabilities as well as for all kinds of achievements in profession, art and culture, which take nowadays place. The extraordinary fortunate developments of our age in some parts of the world, namely in the more happy parts of the western culture, are certainly based on the ability to contain emotivational load and to manage remaining occurrences of this kind thoroughly. In the contemporary western culture, and particularly in professional contexts, it is an absolute must, to keep lower needs, deficiencies and stronger emotivational stress factors under control. This is the inevitable precondition for active participation in the contemporary society. Only people who are able to have the emotivational aspect of life under control and who are able optimise constantly their skills in intellectual, technical, artistic or sportive fields, are able to make benefit of the society, and only people who are well-integrated into the society, so that all needs can be managed sufficiently, namely the lower needs, are able to contain emotivational challenges to the right extent, so that skill optimisation is not impaired too much – this is a mutual conditionality in terms of the question, if one is well-integrated into the society and can keep on-line with the multifarious progressive developments. This is like a demand for coherence on social level, or a great competition in competence, affluence and happiness, which results every day in great achievements.

But there are also many shady sides of this great competition or of this social demand for coherence. Negative consequences can be described in relation to several fields, so for cultural, ecological, economic, political, (socio-) psychological, neural, physiological or other layers of reality. This is a huge topic, but which cannot be faced here in its entirety. So we will keep track of the neural and socio-psychological aspects of life and turn to the most crucial drawback of the social demand for coherence.

A crucial disadvantage or drawback of the *great competition* is, that it tends to challenge people and also teams who cannot keep pace with the progressive developments, to an extraordinary extent; in this connection, term team is meant in the sense of social classes, ethnics, nations, states, regions, families, working groups, companies, institutions or any other groups of people (team means any "social entity"). That these people or *teams* are challenged means that they are confronted in their living environment with significantly higher degrees of emotivational load than tolerable, so that their chances to close the gaps in the informational disciplines, as IP * IB, cognition, intellectual and professional competencies etc., are rather little. It is part of the normal social dynamics that some people and groups are better integrated in one moment and others suboptimal, with the consequence that the latter have significantly more complexity stress and possibly also less benefit. But *normal dynamics* means that there are chances, in terms of a certain grade of integration and equal footing, and that there is a certain probability that the circumstances may develop and the balances may change. The talk can be of a "challenge to an extraordinary extent" if the balances in the mental parameters, so between IP, IB and EB, of an individual or of a team in relation to other individuals or teams, namely the representatives of the *great competition*, is too large, so that they have little chances to come in cooperation with them on a par. The consequence is that they may either live in a niche, not much affected by other people and groups with significantly different mental profiles, so that they can be happy in their living environment (see the north sentinel islanders), or that they are confronted with other, very different facets of the global culture, with the result that additional complexity stress arises due to the large discrepancies in the mental parameters. An extraordinary challenge is provided by the fact that the tension fields between the different cultures make convergence and cooperation the more difficult the larger the gaps in mental boundary conditions, when cooperation attempts are started. Examples: development politics, which fails again and again in many cases; the support and encouragement of children from disadvantaged or poor backgrounds is also often a bigger challenge.

However, one can ask: why is there still larger emotivational load in our world? We, the global humanity, are actually very rich equipped with skills, competencies, material resources and money. We have nowadays (actually, seemingly) much more than we need. Could this not lead to the result that basic needs, deprivation, frustration, shortages widely vanish and that we indulge in the development of growth needs and happiness? There are many reasons why this does not occur to a sustainable extent – this can again be explained in relation to several fields, so for cultural, ecological, economic, political, (socio-) psychological, neural, physiological or other layers of reality, and we will again keep track of the neural and socio-psychological aspects of life. This means that we will now collect explanations, why emotivational load has still so much influence, seen from the perspective of the SPP-4DI model.

The importance of the emotivational dimension is still high or might never vanish because of the following reasons:

- The nearly antagonistic inconsistency in mental profiles between the affluent (western) society and *extraordinarily challenged* people and teams, as described above, is one source of emotivational load and additional 3D/4D complexity, which will stay for a while, and which needs greater attention.
- Each society, even the best, is constantly confronted with erosion processes. Social dynamics and natural processes lead always to deplorable and difficult fates, loss of solutions, structures and competencies as well as additional problems and rising tensions in the society

(between different mental profiles etc.). There are always new challenges to be faced, and the primary mechanism occurs in the nervous system – it is frustration and complexity stress due to new problems, which drive humans firstly crazy and then, hopefully, to the development of inventive solution approaches.

- Higher needs are based to a certain extent on negative experiences, dependency chains on lower needs (see also section "Higher needs" above) and on level specific representations of lower needs on higher need levels (see section "Dynamics in the need hierarchy" above). The result is that lower needs are always involved at each higher need level again, bringing passion, empathy, power and purpose into these need levels, but potentially also adverse effects, like conflict, ruthlessness, violence and war.

- Life is characterised by birth, childhood, adolescence, illness and death to a certain extent. All these phases are a constant source of deficiencies and difficult experiences. But a positive aspect of this is, that negative experiences are the raw material for new skills in two regards: they provide or enforce motivations to overcome the issues, and they are helpful as anti-patterns during exploration and training. Rich amounts of negative experiences are not less valuable than positive skills.

- The social sphere has great importance in society and in the system of needs, and the processes in this sphere are usually to a larger extent based on emotions than on informational communication. The social need level 3 may temporarily take the role of the most challenging need level under affluent circumstances, while there are few reasons for the deprivation at the lower need levels 1 and 2. But social dynamics can also result in extraordinary large amounts of emotivational load, not least because of dependency chains on lower needs, such as e.g. the need for living wage and accommodation (level 2) and sexuality (level 1). Examples: dismissal due to a conflict in the working group, divorce/separation from partner; both types of events are loaded with serious emotional impacts.

- The implications of the phenomena "Emotivational amplification adaptation" and "Fading consciousness in affluent contexts" as described in the appropriate sections below. These mechanisms result regularly again in higher amounts of emotivational load, although if stronger emotions, deficiency needs and larger levels of stress could actually have been managed in consistent manner on the basis of rich competencies and resources.

- In abstract terms: reality cannot be lost; one will always be grounded again by additional difficulties if floating in higher spheres for a while. It can be possible to skip this rule for certain parts of the material/external aspect of the need system − by extraordinary riches −, but for the aesthetic side, the brain processes, healthiness and social relationships it might not.

The tension fields and dynamics in the 3D/4D space are primarily caused by difficult circumstances, deficiencies and shortages as well as by appropriate mechanisms in brain processes, as pain, homeostatic imbalance, emotion, frustration and all kinds of basic needs or deficiency needs. So it is no wonder that they work well under these circumstances, according to the findings from the sections above. But the message of the explanations in this section is, that exactly these mechanisms take also effect under circumstances of pronounced competencies, affluence and happiness, not least because there is no other set of motivation mechanisms available in the brain processor. The tensions and stress levels can still be very high. Under better circumstances, the only difference is that the tensions occur more in relation to sophisticated artificial or growth needs, which are maintained with the behalf of the reward system, than in relation to the more stringent deficiency needs.

The tunnel vision paradox

A consequence of the "Dynamics in the 3D tension field" and the "Primary compensation mechanism in the 3D space" (see figure 3) is also a phenomenon that can be called "tunnel vision paradox". The primary or crucial compensation mechanism in the 3D space is provided by the typical reaction to incidents of pain, frustration or stress. Due to such kind of incident, it is an involuntary mechanism that the brain processes focus on important patterns and fade out other information – see the IB curve in figure 3. The self-regulation system, which is implemented in the brain and adjacent systems (autonomic nervous system, endocrine system, body), works then typically in a regulatory way which leads again to the reduction of the pain or of the emotivational stress level (EB), with the demanded result, that the former informational scope can be re-established (IB).

But that the issues can be solved in this form, is not, like the *primary compensation mechanism*, an obligatory modus operandi. Particularly in connection with the larger complexity and proceeding sophistication in this day and age, it can also happen that no simple or direct solution is known, so that detours are necessary to come to a satisfying result. This means that additional information must be acquired, skills must be learned or that further analysis, research and development must be done for the ability to approach a solution.

The consequence is, that one must look around for information and suggestions which could help to solve the problem. But this contradicts the fact that lowered informational bandwidth is an inevitable consequence of pain and increasing problem stress. This dilemma is exactly what can be called "tunnel vision paradox": the capability to overcome *focusing* and *tunnel vision* is the crucial key to find a solution, but the problem stress, which has to be solved, causes and intensifies these phenomena at the same time, so that the solution process is hindered in highly effective manner by the corresponding problem. The curves in figure 4 show this dilemma – the actual desire is to get the emotivational load decreased (dotted line at the EB curve) with the result that the degree of informational focusing can be relaxed (dotted line at the IB curve). But this does not work in many cases, so that it is required to defocus despite the perceived pain (dashed line at the IB curve) for the ability to achieve a solution.

A very simple example is provided by a specific free-climbing skill – the capability to use the legs to relieve the hands. On a naive perception, free climbing is a series of pull-ups, to be performed by the muscle power of arms, hands and fingers. But this view is widely wrong, and the real truth behind free climbing comprises much more. One point is the right usage of legs and feet.

During a climb, it is a typical scenario that the power resources fade away and falling down becomes imminent. This is always more or less scary, and it is a normal reflex that one looks for the next grip above which is hopefully rough enough, so that it can be hold with the remaining finger force reserves. But it is a typical challenge that this is does not work easily in some cases – there might be structures, but they are too small or too slippery for being entered with pure finger and arm power. In some cases, the solution is, to place a foot on the next small edge, what results in a gravity compensation effect for body and hand, so that the next

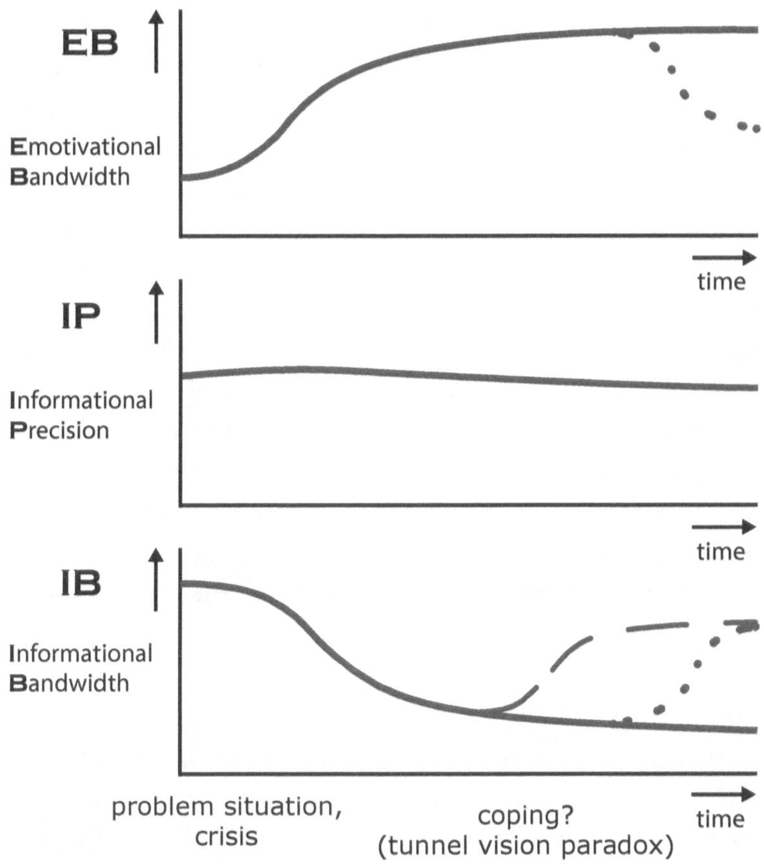

Figure 4 – The tunnel vision paradox in the 3D space

small crimp can now sufficiently be hold with the remaining finger power.

Even if this is theoretically already clear, what it is often not the case – the crucial challenge is always provided by the tension fields in the brain. Simply grasping upwards is a natural reflex in an overhanging climbing wall, which works in many cases properly, but not when the climbing route gets more difficult. The normal result is increasing struggle and fear and, at the same time, fading power reserves. There is no easy way out of this situation – due to the *tunnel vision paradox*. There is rather an urge to strengthen the mental focus, letting no other choice than grasping upwards with more power than before. But this is a terrific standoff in some cases, which leads to no other result than losing the last power reserves and falling in the end.

The error is that the eyes have not been turned downwards or that it was not possible to remember the wall structure which has been visited some seconds before, when the last climbing move had been done. That's why the small foot hold has not been discovered, which could bring salvation. But even if this edge is detected, it is not obvious, that it will help. The climber needs also enough coolness or experience for being able to develop the required imagination – that it could be helpful to enter this edge with the foot before the next grasp attempt. This is a learning process, which is naturally blocked by panic-like fears or even by the *tunnel vision paradox*, respectively.

But it is part of the human fate that he has to overcome such kinds of mental standoffs. For the free climber, this means, that he might try this wall again and again, until the required climbing move could be developed and executed successfully. And it might be typical for the *tunnel vision paradox* that it is not overcome by facing directly towards the problem, but rather via detours, which are more or less complicated. In the case of the free climber, this means that the training equipment is sufficient, including rope and harness etc., that the bolts in the climbing wall are safe, that he has a good rope team, and that he may think about the solution in intervals, particularly when he is not directly under pressure, so that the antagonistic tunnel-vision blocking is temporarily released. Thus, switching between different contexts might be the best way to come to solutions and to realise the dashed line at the IB curve in figure 4, and consequently also the dotted line at the EB curve.

The *tunnel vision paradox* works like an antagonistic inconsistency, in the sense that most of our problems block always at the same time their solution. And the capability to overcome this antagonism again and again is the most important skill in this day and age. But this is never easy, it is rather always a process which is more or less complicated and which may often need days, years, generations or ages and major efforts, until a real good solution can be achieved in the end. And some problems will never be solved, or tend always to escalate again. There are always endless concrete explanations why problems cannot be solved. But the *tunnel vision paradox* should be an abstract explanation, which is in each case part of the truth. So it might also provide suggestions which could be helpful in general, what could be a topic for another chapter or for another book.

Emotivational amplification adaptation

Even if all lower needs (at levels 1-3) can be contained and excluded to a great extent, emotivational dynamics may take effect with high amplitude. There is no motivation and no aim without certain differences between goals and current achievements. This is the crucial mechanism for brain activity; brain processes would cease and completely vanish if there are no differences between desires and current states. Thus, it is inevitable that remaining deficits in growth ambitions, even if they are very small, are amplified until they are large enough to provide strong motivations. Vice versa, sophisticated differences are naturally fading, when stronger impacts occur again in the brain processes. This is a basic neural regulation mechanism which might also be described as "trend towards subtlety". Term "subtlety" means in this case, that very subtle or sophisticated differences – as e.g. a foot position on the dance floor, small shades of cloth colours or special formulations in a letter – are made extraordinary important matters, which can involve significant emotivational incidents in the brain processes. This is necessary because self-regulation would be undermined and the evolutionary determination would be lost in affluent contexts otherwise, but what is not possible due to human nature. A kind of emotivational amplification adaptation occurs in our brain, which is able to make major incidents of small sophisticated details that are actually nonsense, seen from a more existential point of view. This mechanism works precisely over the projection of emotivational signals to the informational complexity which is applied along with the sophisticated matters and by taking them extraordinary serious by this way.

The perception of emotivational load occurs not per se or in an absolute or fixed manner, but it enfolds rather towards informational complexity. A simple example with a basic need: The same degree of thirst, in absolute terms, might be perceived with different strength in the following scenarios: scenario 1: The next grocery store is seen at the end of the road, the purse is filled; scenario 2: the purse is empty; the addresses of the next cash point and grocery store are known, but the locations can only be reached via a complex route (note: the grocery store is known to only accept cash in both scenarios). Even under the assumption that the time to get the thirst quenched is equal for both scenarios, scenario 2 might be perceived as much more painful because of the informational complexity and the higher degree of uncertainty, which has to be managed. Thus, the degree of the perceived emotivation is implicitly the higher the more informational aspects and details are involved at the same time. Another example with a social need (need level 3), but which may also involve lower need levels via dependency chains, as e.g. the need for accommodation (level 2) and sexuality (level 1): The marital life of a woman and a man is impaired by a domestic quarrel, but both people have to go to work; scenario 3: the productivity of the man, who is working as mathematician, is significantly impaired because he can hardly concentrate on a complex topic, which involves many issues and stakeholders, and which requires an inventive solution by a smart algorithm; scenario 4: the women's productivity is not affected much – she makes a routine job for a facility cleaning service.

The book that emotivational load enfolds towards or is amplified by informational complexity is also consistent to the mental capacity formula (MC) and to the adaptation mechanisms which have been described in the section above. Focusing and fading out of too much aspects in the IB dimension, or also of too much details in the IP dimension, is not least an immediate and inevitable reaction, because of the rule that information takes effect as amplifier of emotivational pain. While the mental capacity is seen as fully exploited in scenario 2 from above or as overloaded in scenario 3, it might have remaining shares for thinking about additional issues in scenarios 1 and 4. The consequence is that emotivational forces can take effect with similar vigour, if either driven by strong primary emotivational signals in combination with simple information or by weak primary emotivational signals in combination with high informational complexity. That's why in affluent contexts life gets extraordinary complex and motivations are still very high.

Fading consciousness in affluent contexts

Consciousness has been explained as *information enlightened by emotivational evaluation.* This is seen as based on projection between information that is perceived from an "external" focal plane, which is mainly provided by the *environment* (but see also the hints below), on the one hand, and emotivations that are elicited from an internal focal plane, which is provided by the organism and its physiological regulation processes (by the *body*), on the other hand. This projection occurs at the *evaluative focal plane* in the brain, namely in the so called attention assessment controller (AAC). A special point in this connection is that some systems, so somatosensory perception, elicit both information (… about movement of limbs and body) and emotivation (pain); consequences are that different outcomes from this system are incorporated at the evaluative focal plane from both the informational and the emotivational side and that information comes not only from the environment (external), but also from the body (internal; but it is indeed information about interactions with the environment in terms of another sense). Another special point is that present informational-emotivational patterns are continuously stored in the memory, with the effect that both components are later involved again in the processes at the evaluative focal plane, so that consciousness has always to do with memories. Another special point is that evaluation occurs in conscious and unconscious manner separately for each of the two components – information and emotivation; consciousness is usually only attributed to the conscious part of the processes, but while the unconscious part must be seen as the raw material for this, so that it can also be seen as involved.

Social consciousness has been explained as individual consciousness, but occurring for many people, mediated via the so called *social focal plane*, which is provided by the social interaction and communication processes.

So far a short summary about what consciousness might be. More details see above in sections "Conscious experiences", "The two types of consciousness" and "Individual and social consciousness".

However, an important message of this is that consciousness occurs as combination between information and emotivation. That's why the degree of consciousness is the larger the higher the quantity of both components.

Another aspect of consciousness is the relationship between deficiency needs and artificial/growth needs or the relationship between the different need levels (1-5). It has been outlined above that the development of artificial needs or growth needs is a typical strategy of the human to manage his affairs under the conditions of evolutionary competition of species in sustainable manner. But at the same time it is clear that artificial needs, so the needs at levels 3-5, do not end in themselves – they are rather only a more or less complicated detour to ensure that more basic needs are satisfied in sustainable manner. Of course, the neural pattern library contents and the brain processes of the human, including experiences, memories, skills, evaluation processes and control processes, can be characterised as the more comprehensive the better they are developed in relation to distinctive artificial needs (and thus cultural aspects). But the other side of the coin is that the higher-order brain processes and brain contents are founded on basic experiences and skills which occur in relation to more fundamental needs – this is always the starting point, raw material and final purpose of higher-order ambitions. This is also supported by the findings from above that higher needs may develop to a great extent on the basis of *emotivational dependency chains* and *emotivational stimulus modulation* (see section "Higher needs") and that lower needs get often representations on higher need levels (see section "Dynamics of the need hierarchy").

The relationship between the different need levels corresponds with the relationship between information and emotivation insofar as that information, particularly informational complexity, is naturally associated with higher need levels – in intellectual, artistic, economic and cultural spheres –, and that emotivation refers to the existential side of life, and thus to prevailing basic needs, forcing emotivations, focusing and the necessity to reduce informational complexity.

So it should be clear that a certain balance between information and emotivation, as well as between deficiency needs and artificial needs should be the basis of spirited brain processes and of consciousness.

That consciousness is not very high without a certain degree of intellectuality or neural-informational complexity, is rather totally clear. Nobody might deny that the intelligence and consciousness of humans is greater than the intelligence and consciousness of primates who have few intellectual/ artistic/ economical/ cultural ambitions and who develop few artificial needs. This view requires no defence.

But what about the contrary aspect? What happens if informational complexity is provided in rich amounts and emotivation tends to fade? Under affluent conditions, it is a natural general ambition to get away from deficiencies and from the world of existential experiences. And many of us are successful in managing this ambition. But what is the consequence of this type of success?

The book is here that this leads again to fading consciousness, and consequently also of intelligence!

If the world of artificial and growth needs prospers like a wonderful orchid and basic needs are only known in terms of their representations at the highest levels, so as aesthetic experiences and narratives, but not in terms of real threats, for a long time, the result should be that the higher-order ambitions make themselves independent from the existential side of life. This leads to loss of reality and to a nonsense culture, which is decoupled from our evolutionary ground. This is a special kind of unconsciousness, which is caused by emotivational randomness rather than by a lack of informational complexity.

Fortunately, this is only a tendency, and there are many mechanisms which bring us back down to earth. There are always also existentially challenged people and teams in the world – in the sense of social classes, ethnics, nations, states, regions, families, working groups, companies, institutions etc. –, who have bigger problems and need support, there are erosion processes, birth, childhood, adolescence, illness, death and social disasters which ensure that emotivational load never ends, and it is clear that temporary loss of reality results always in additional problems and thus in reinforcement of the emotivational factor as ultimate principle (see also section "3D tensions in the affluent society"). So it is clear that there are enough forces to let us not lose our evolutionary ground.

But this happens not only by chance and by the back-down-to-earth mechanisms from above. Many of us are obviously able to perceive fading consciousness, caused by greater intellectual and material affluence or by far-fetched skills and too much comfort, as threat against sensible reason. That's why they search for more fundamental challenges.

Examples of this might be the following:
- Carriers – even of very talented people – turn often from sophisticated scientific topics to more pragmatic professional fields.
- Challenges with physical or even existential component are getting increasingly popular, such as climbing, flying,

diving or any other types of sport.

- Many people seek fulfilment in charitable ambitions, with the result that they establish connections from relatively affluent contexts to more difficult living conditions.

So it is always ensured in any form, either by natural laws or by active ambitions, that consciousness and sensible reason is not fading too much. But it is important to understand this aspect and to know that arbitrariness, whateverism, decadence and loss of control due to fading consciousness due to loss of emotivational fundamentals might be a typical societal problem in this day and age.

About the integrative ingredient of 4DI

Intelligence of the 4DI type could also be named "integrative 4D intelligence" (I4DI). There are several reasons for this.

First, intelligence is based on integrative encoding in the declarative association areas of the brain, so in the long-term memory: "Integrative encoding proposes that when a new event shares a common feature, or features, with an existing memory trace, the common features elicit memory reinstatement through hippocampal pattern completion. New experiences would then not only be encoded in the context of presently available information in the external world, but would also be bound to reactivated representations of prior related memories." See in Davachi and Preston 2014, 544, and section "Integrative encoding of memories over time" in chapter "Emotion, motivation and memory" in Part 2. This includes also sensorimotor patterns from the procedural memory, which are integrated in the declarative memory in the form of microrepresentations, as well as emotivational patterns, which are integrated as coherent value ingredient in key areas of the declarative part of the memory system. See also section "Integrative encoding of procedural, emotional-motivational, sensory-perceptual and internal thought components" in chapter "Emotion, motivation and memory" in Part 2.

Second, consciousness is based on the capability to produce maxima of integrated information and to differ this information from other integrated information. Conscious experiences occur only if an information processing system achieves a certain degree of complexity in this regard. See also section "Tononi, G.: The integrated information theory of consciousness" in chapter "Consciousness" in Part 2.

Third, it is a central principle of the 4DI concept, and hence of the brain, that complexity is faced in the tension field between informational bandwidth, informational precision and emotivational bandwidth, and that this challenge is managed successfully. This is nothing else than a great integration capability in an ever-changing environment with multifarious aspects (however, naturally by considering the own stakes and the stakes of the species as major motivations).

Fourth, involves dimension (4) – team intelligence – social integration with no limit.

Toe-holds for other disciplines

Might a general model of the brain, as, e.g., the SPP model or the SPP-4DI model, provide special neuroscientific toeholds for other disciplines, like psychology, social sciences or politics?

Yes, both the SPP model and the SPP-4DI model might provide special toeholds, which could be helpful for interpret psychological, social or political phenomena from neuroscientific perspectives. The four disciplines can be brought in a closer relationship than before, albeit only on a hypothetical basis. (Note: the political science is also seen as part of the social science, so it could be more correct to refer to three disciplines rather than to four.)

Examples are the following:

- The concepts "two types of consciousness" and "4DI" provide a better (hypothetical) explanation of the interdependency between rationality and emotionality than ever available before.
- The interdependency between the emotivational factor (EB) and the informational factors (IP, IB) in the neural control system, and the mental capacity formula (MC = IP * IB * EB = constant) might be a key to many psychological phenomena. This might be valid for normal reactions to pain and fear, mental diseases – like traumata, depression, burnout, dissociative phenomena –, as well as for the usual dealing with daily challenges – like exam situations, problem solution processes, project management, crisis management, political processes etc.
- Humans are much better adaptable to ever-changing environments than animals. But there is a remaining level of clumsiness, and new solution approaches are often only adopted after deeper crises and via complicated de-

tours rather than on a pure analytical or rational way. The "tunnel vision paradox" can basically explain why, but further investigation of this matter should be necessary to come to a certain understanding of the appropriate mechanisms.

- The SPP-4DI concept can be a good theory which explains the way of life and the development of the global human society. Good local and global politics, and especially a beneficial dynamic between both aspects, is just a complexity issue, and the 4DI concept can be a good starting point for investigate of how humans, as social beings, might be able to deal with growing complexity and with crisis situations. The 4DI concept might also explain why crises occur again and again, and how this aspect of life could be managed in a better way.

- The yin and yang between informational complexity, sensitivity and carefulness, and thus IP * IB, on the one hand and kinds of mental strength, power and crudeness, and thus EB, on the other hand might provide better explanations of the balances between political movements than any other theories. This might also explain of how times of affluence and crisis or times of peace and war alternate in cyclic manner.

- The capability of the human to stop inappropriate activities and habitual behaviours immediately and search for alternatives, which is also called "impulse control" or "preparation mode" or "control level 4/5", has developed by the evolution more in relation to the societal context rather than in relation to the natural environment; the natural environment was traditionally the general opponent, which was to be combated, tackled and exploited on the basis of social cohesion. This might among other things explain why the contemporary society, as a whole, remains on the path of resource-intensive growth, although there are strong indications that this will lead into the ruin of the natural environment. The principle is simple: social integrity and thus continued progress on the way once chosen matters, integrity of the natural environment is less important. This approach was actually not wrong, since the natural environment was originally able to take care for its own matters. But this era has changed into the opposite, and human intelligence, which is based on the impulse control capability, is maladapted to this new balance of power.

- That the discrepancies between western cultures and "the African way of life" are difficult to solve, may not least be caused by major differences in the mental 4DI parameters on sociocultural level. Intelligence might not be higher or lower under different circumstances, but it works in different ways. That there are still many poor people in affluent countries might have a similar cause. See also section "3D tensions in the affluent society".

A crucial question might be: How deals the brain with complex real-life challenges in this day and age? The SPP-4DI model can provide suitable hints to come to an answer:

The brain is able to internalise a world model. This occurs particularly in the associative-emotivational part of the brain, so in the declarative memory, but accompanied by further optimisations in all other parts of the brain, namely also in the sensorimotor-emotivational part, so in the procedural memory. These neural solution pattern libraries are constantly optimised and exploited in real-life situations for the sake of the individual and the species.

It is clear that the sufficiency and *suitability* of this system must always be limited for several reasons. But the most crucial limitation is not the interconnection complexity which can be achieved in general – the brain should be a scalable organ, which may significantly grow over generations –, the book is rather that the most crucial limitation is provided by the limited ability to suffer pain and dissatisfaction. This is kind of a vicious circle: Pain and dissatisfaction take at the same time effect as goal providers and also as signals that are counterproductive for interconnection complexity. Conversely, opens decreasing pain not only the way to internalise higher environmental complexity, but at the same time also to lose its relevance and sufficiency gradually. This kind of loss of reality leads reliably again – via crises (or even wars) – to increasing pain and dissatisfaction, and consequently to the down-regulation of interconnection complexity. The capability to overcome this vicious circle step by step correlates with the ability to manage real-life challenges increasingly in controlled manner by the development of mental capacity (MC) in the sense of the 4DI model (see also section "Dynamics in the 3D tension field").

Another important question is also: What is the actual strength of the human and particularly of his central control apparatus which is provided by the brain?

It is not skills or cognition or the ability to develop scientific knowledge! It is rather the capability to develop systems of artificial needs in a way so that this releases success and joy during the behavioural detours which lead also to the fulfilment of all other more stringent demands in the end. Skills and knowledge are just executive components, which could not be missed, of course, but their development is driven by the primary capability from above. And the development of tools and the exploitation of material resources are just tertiary aspects of this primary duality.

Part 2 – Excursions to the current state of science

Introduction

This part summarises some insights and conclusions from external references. It is a loose collection of discussions about several aspects of the brain. There is no systematic or purpose behind these descriptions other than providing pre-compiled findings for Part 1 of the book. All sections in Part 2 are involved into the concepts from Part 1, what is always highlighted there by references to the corresponding sections in Part 2.

Basal ganglia (BG) and frontal cortex

Introduction into the BG matter

The subcortical area which is called basal ganglia plays an important role in the brain. It represents affective decision making (in the ventral striatum) and is involved in processes like motor control, control of associative/cognitive processes and emotion control, only to name the most important functions. It is interconnected, among other things, with the motor cortex, somatosensory nerve cords, posterior parietal cortex – representing multisensory integration –, with the dorsolateral prefrontal cortex and other prefrontal regions – representing executive functions, cognitive and associative processes, working memory, abstract reasoning, planning and value-guided decision making; see also section "Decision making in the frontal cortex" –, and it is, respectively, part of or integrated with the limbic system – representing homeostasis, emotions, reward system, goals, motivations.

According to Gazzanega and Mangun 2014 and articles and references therein, there are indications that the basal ganglia provide the following performance features in collaboration with adjacent and interconnected areas:

- Affective decision making, closely integrated with value-guided decision making in the prefrontal cortex (see section "Decision making in BG and frontal cortex" ff.).
- Control of motor sequences and stimulus-response associations; parallel management and mediation of several brain functions via distinct control circuits, comprising at least "skeletomotor, oculomotor, frontal associative, and limbic functions" (see section "The four parallel circuits through the basal ganglia (BG)").

- Impulse control/response inhibition for "potentially interfering or inappropriate actions", suppression or facilitation of movements, reinforcement driven learning (see section "The motor circuit").
- Regulation of movement vigour, in terms of movement speed, amplitude or force (see section "The motor vigor hypothesis").
- Control and modulation of language (see section "Basal ganglia (BG) and language").

The four parallel circuits through the basal ganglia (BG)

Four "anatomically and functionally distinct parallel circuits" are described in Turner and Pasquereau 2014 as follows:

"The basal ganglia (BG) is composed of several heavily interconnected nuclei at the base of the cerebrum. A series of anatomically distinct parallel circuits through the BG receive afferent projections from and project back to cortical regions that mediate skeletomotor, oculomotor, frontal associative, and limbic functions." (435)

"Anatomically segregated parallel circuits through the BG subserve skeletomotor, oculomotor, associative, and limbic functions. All circuits obey a common internal organization. [...] BG-receiving subregions of thalamus project back to a subset of the cortical areas that project into each circuit." (436)

These four circuits or loops, respectively, are interconnected as follows:
- the skeletomotor circuit with the motor cortex, controlling sensorimotor functions,
- the oculomotor circuit with the frontal eye cortex, controlling eye movements,

- the associative cortex with the dorsolateral prefrontal cortex, representing cognitive functions and
- the limbic circuit with the anterior cingulate cortex, representing emotions.

It is asserted that there are further loop circuits that include temporal and parietal cortical areas as well as thalamic nuclei.

All these circuits are not to be seen as strictly segregated, they are rather integrated by some kinds of "projection", "communication", "convergence" and "funneling". Pathways are mentioned, for example, which support projections from limbic regions to nonlimbic frontal cortical areas or to motor cortices.

(See Turner and Pasquereau 2014 and references therein.)

The motor circuit

Turner and Pasquereau 2014 focus primarily on the motor circuit, starting with headline "Organization and physiology of the motor circuit" (437ff.). They emphasise some special roles of this circuit under headline "Functions of the basal ganglia using the motor circuit as a model" (443ff.):

"(1) the suppression or facilitation of movements",

"(2) reinforcement-driven learning, especially of habits or skills" and

"(3) the regulation of movement vigor".

Functions (1) and (2) mean "that the BG motor circuit contributes to movement selection or initiation" and "that the motor circuit is involved in retaining or executing well-learned motor skills like motor sequences or stimulus-response associations". (445)

More sophisticated findings on skill learning, described at pages 443f. and 445, suggest that it is in fact an important function of the motor circuit to suppress inappropriate actions and develop more suitable actions, but that well learned actions and habits work usually better without the interventions of this basal ganglia circuit.

Based on several studies, Turner and Pasquereau (2014) come to the following conclusion:

> "Thus, multiple lines of evidence indicate that the BG promotes new skill learning, but that other parts of the brain (cortex in particular) take over the storage and production of well-practiced skills. The unique neuro-modulatory milieu of the striatum provides an ideal substrate for rapid reinforcement-driven plasticity, but cortex is better suited for long-term retention and execution." (445)
>
> (See Turner and Pasquereau 2014 and references therein.)
>
> Notes: BG means basal ganglia; the striatum is the region in the BG which represents affective and value-guided decision making (see also below in section "Decision making in BG and frontal cortex").

Further descriptions in Turner and Pasquereau 2014 support the finding that a role of the basal ganglia is to act as intervention switch, which inhibits inappropriate responses or actions, rather than to support well-trained skills and behaviours; that inappropriate responses can be stopped very quickly and perfectly seems to be an extraordinary important feature, so that a special "hyperdirect pathway" has been developed for this purpose by the Evolution; this role of the basal ganglia might be particularly important in social contexts while, for well-trained skills and behaviours, the intervention function of the basal ganglia is identified more as a disturbing than an adjuvant factor. Among other things, Turner and Pasquereau come to the conclusion that the "the cortico-STN-GPi pathway may play a key role in societally important impulse control disorders, such as those associated with compulsive gambling and substance abuse".

(See Turner and Pasquereau 2014 and references therein, particularly the descriptions at page 442 under headline "IMPULSE CONTROL DISORDERS AND THE CORTICOSUBTHALAMIC NUCLEUS 'HYPERDIRECT' PATHWAY" and at page 440f. under headline "Clinical conditions associated with basal ganglia dysfunction".)

Notes: STN means subthalamic nucleus; GPi means internal globus pallidus; both regions are seen as part of the basal ganglia.

Motor vigor hypothesis

According to the so called "motor vigor hypothesis" (Turner and Pasquereau 2014, 445), function (3) "the regulation of movement vigor" (see above) means that "the motor circuit regulates a form of implicit 'motor motivation' that influences how vigorously individual actions are performed", what includes "movement speed, amplitude, or force" (Turner and Pasquereau 2014, 445).

An open question is asked as follows: "It remains to be determined if the vigor idea applies equally to all functional circuits of the BG (i.e., that each BG functional circuit links motivation to its own form of action) or if it is rather a specific function of the BG skeletomotor circuit." (445)

(See Turner and Pasquereau 2014 and references therein.)

Note: BG means basal ganglia.

Decision making in BG and frontal cortex

The brain is equipped with a decision making system or a "decision making apparatus", comprising functions like "affective (impulsive) decision making" and "value-guided decision making". Taking into account the descriptions in Gazzanega/Mangun 2014, this system can be attributed to two specific regions in the brain:
- the basal ganglia, particularly the ventral striatum and
- the prefrontal cortex, particularly the ventromedial prefrontal cortex (vmPFC) and the medial orbitofrontal cortex (mOFC).

The (ventral) striatum is typically connected with "affective decision making" and the prefrontal cortex with "value-guided decision making". But latest studies suggest that both types of decision making are supported by both regions in close cooperation.

The integrated mode of operation in both regions will be highlighted below (at the end of this section), before the role of the prefrontal cortex is described in more detail under "Decision making in the frontal cortex".

General findings on the topic "decision making" are described in Gazzanega and Mangun 2014 in the following chapters:
- 3 "Cognitive Control and Affective Decision Making in Childhood and Adolescence" (Crone 2014, 23ff.)
- 59 "The Emerging Standard Model of the Human Decision-Making Apparatus" (Glimcher 2014, 683ff.)
- 91 "Neuroeconomics" (Huettel 2014, 1063ff.)

The decision making system is often tested with behaviours in relation to potential rewards, as, e.g., described by Crone 2014 and references therein under headline "The neurocog-

nitive development of affective decision making" (26f.) at
page 27:

> "SHORT-TERM AND LONG-TERM CONSE-
> QUENCES The way individuals weigh short- versus
> long-term consequences is often examined in delay
> discounting tasks where the choice for an immediate
> reward (e.g., 1 euro now) and a delayed reward (e.g., 2
> euros later) is presented with variable delays. Individu-
> als tend to prefer the immediate reward more when the
> delay for the larger reward is longer. In a delay discount-
> ing task, preference for a delayed reward was associated
> with activation in the lateral prefrontal cortex, whereas
> preference for an immediate reward was associated with
> activation in the ventral striatum in both adolescents and
> adults".
> Note: The (ventral) striatum is a part of the basal gan-
> glia.

But this kind of clear division of functions between affective (impulsive) decision making in the ventral striatum and
value-guided decision making in the prefrontal cortex is
questioned by Huettel 2014 and references therein at pages
1065f.:

> "TEMPORAL DISCOUNTING Everyday decisions
> often involve a consideration of future consequences.
> When deciding to save for retirement, people forego
> smaller rewards now in order to obtain larger rewards at
> a later date. [...] An early and very influential study [...]
> posited that intertemporal decisions arise from the com-
> petition between two competing neural systems: a rel-
> atively impulsive system involving value-related regions
> (e.g., ventral striatum and OFC) and a relatively patient
> system involving control-related regions (e.g., dlPFC).
> [...] There is a growing consensus, accordingly, that the
> brain contains a core system for processing value – not

two interacting systems – but that factors like the exertion of self-control can shape processing in that value system."

Abbreviations: OFC: orbitofrontal cortex; dlPFC = dorsolateral prefrontal cortex.

That decision making occurs in a highly integrated system, which is composed of the striatum (in the basal ganglia) and parts of the prefrontal cortex, is also supported by Glimcher 2014 and references therein at page 686:

"THE VALUATION CIRCUIT [...] one way to begin to understand valuation circuits was through the study of dopamine. At the same time [...] a second line of inquiry with regard to value also got underway: the search for functional MRI (fMRI) signals that correlated with the subjective values of goods, actions, and events expressed in various ways by human subjects [...] Both of these approaches converged on two particular brain areas: the striatum and a portion of the medial prefrontal cortex. Activity in these two areas was found to consistently predict people's preferences – and preferences of literally *all* kinds. If someone was a *steep temporal discounter*, she preferred small immediate rewards to waiting for larger rewards, and activity in these areas in response to delayed rewards was steeply 'discounted' [...] If someone placed a higher value on food rewards than on monetary rewards, then so did these areas in that person [...] If someone was inclined to co-operate in a trust game, then activity in these areas was higher for offers that required cooperation".

Abbreviations: MRI: magnetic resonance imaging.

Decision making in the frontal cortex

The prefrontal cortex is connected with executive functions, cognitive and associative processes, working memory, abstract reasoning, planning and value-guided decision making. The following descriptions will concentrate on the topic "value-guided decision making", since this might be a key function for all other performance features which are linked with this region.

This topic is addressed in Rushworth et al. 2014 and references therein. The abstract of this article reveals already much:

> "Actions are chosen on the basis of expectations about the benefits they will yield. Areas in the brain's frontal lobes are important for making decisions about which choice it is best to take. In functional magnetic resonance imaging experiments, the ventromedial prefrontal cortex exhibits signals related to the values of potential choices, and the activity pattern is consistent with the emergence of a decision. Ventromedial prefrontal cortex lesions disrupt value-guided decision making. Another frontal lobe area, the anterior cingulate cortex, is also important for value-guided decision making. There is some evidence suggesting that anterior cingulate cortex is especially involved in the selection of an action in the light of the rewards that will follow and the effort that will be invested. More fundamentally, it may be concerned with the computation and comparison of values needed for decision making during foraging. During foraging, the key decision is whether to engage and pursue the default choice, or whether the foraging opportunities available elsewhere in the environment mean it is better to switch away from the current behavior and pursue other alternatives."

The following findings are described in this article, among other things:

- "The ventromedial prefrontal cortex (vmPFC) and adjacent medial orbitofrontal cortex (mOFC) play a central role in choice selection." (501) (vmPFC: ventromedial prefrontal cortex; mOFC: medial orbitofrontal cortex)
- Important parameters are "reward magnitude (size)" and "reward probability". (505)
- The decision making procedure comprises two parts, "the selection of a reward goal that becomes the focus of behaviour" and "Once the reward goal is selected, then a second type of decision has to be made about which action should be made to obtain the reward goal." (505)
- Decisions might have to be done in terms of binary decisions between two rewards (best option, second-best option), but it is more common that they occur as search/engage decisions – e.g. about "whether to engage with a potential food option or whether to carryon searching for better opportunities elsewhere" (506). Reports about investigations on both types of decisions are comprised in Rushworth et al. 2014.

That the ventromedial prefrontal cortex (vmPFC) and the adjacent medial orbitofrontal cortex (mOFC) play a central role in choice selection is evident by some studies as, e.g. described under headline "Disrupting vmPFC/mOFC impairs value-guided decision making" at page 504:

"Macaques with vmPFC/mOFC lesions were significantly worse when the second-best option was close in value to the best option. In other words, vmPFC/mOFC lesions made macaques worse at taking the most difficult value-guided decisions. Moreover, the disruptive impact on difficult decisions was specific to vmPFC/mOFC lesions and did not occur after lesions in other frontal brain areas that contain neurons with reward-related activity patterns, such as anterior cingulate cortex (ACC) or lateral orbital frontal cortex (lOFC)".

But that the two regions vmPFC/mOFC are only responsible for decisions about the selection of a reward goal (also in terms of attention) while decisions about the consequent actions are made in the anterior cingulate cortex (ACC), is explained under headline "Translating values into actions: Integration across frontal lobe systems" at page 505:

"When vmPFC/mOFC makes a decision, it does not do so in the sense of choosing a specific action, but instead the type of decision that it makes is the selection of a reward goal that becomes the focus of behavior. This appears to be especially the case when there are multiple competing alternative choices and attention must be directed while critical comparisons are made between the various options […] Once the reward goal is selected, then a second type of decision has to be made about which action should be made to obtain the reward goal."

"The ACC region that carried both chosen value and action direction signals is well placed to influence the motor system."

"In summary, one way in which different frontal lobe areas might interact during decision making is for vmPFC/mOFC and possibly adjacent OFC areas to make a choice between different possible reward goals and for ACC to decide between different possible actions."

That not only binary decisions have to be made, but rather also "search/engage" decisions, is highlighted under headline "Distinct mechanisms for decision making in ACC and vmPFC/mOFC" at pages 506ff:

"It might be only on some occasions that a macaque foraging in the wild is given the opportunity to make the type of choice that the vmPFC/mOFC appears to make

– for example, a choice between an apple and an orange. Instead, what is more likely to happen is that the macaque might first see an opportunity to pursue a course of action that might lead to one piece of fruit and only later perceive the opportunity to pursue another piece of fruit by taking another course of action. The critical choice for the foraging animal is, therefore, whether to engage with a potential food option or whether to carry-on searching for better opportunities elsewhere."

The assumption that decisions of the search/engage type are possibly processed in the ACC region rather than in the regions vmPFC/mOFC is promoted under headline "Reinterpreting the role of ACC" at page 508:

"The human ACC region that carries search/engage signals for foraging is well placed to influence whether or not a successful action will be repeated or whether the value of an alternative action will be explored." (508)

Some important aspects are summarized under headline "Conclusion" at page 510f.:

"We have emphasized the different functional contributions of frontal lobe regions to value-guided decision making. We have emphasized a flexible vmPFC/ mOFC system that makes comparisons between choice values by focusing on the most relevant aspects of value (for example, reward probability or reward magnitude). In some cases such a system may work in series with brain areas such as the ACC, which might select an action compatible with the reward goal that is the focus of attention. In some situations, however, the ACC might operate in a quite independent manner to take simple stay-switch decisions on the basis of different types of value information such as average search values and effort costs."

Basal ganglia (BG) and language

Chan, Ryan and Bever (2013) investigate the function of the basal ganglia (BG) in language control and come to the following conclusion (292):

"In sum, our study clearly demonstrated that the BG work in conjunction with frontal cortical regions to subserve a language-specific role in sequencing."

Their approach is explained in the abstract (283) as follows:

"The basal ganglia (BG) have long been associated with cognitive control, and it is widely accepted that they also subserve an indirect, control role in language. Nevertheless, it cannot be completely ruled out that the BG may be involved in language in some domain-specific manner. The present study aimed to investigate one type of cognitive control – sequencing, a function that has long been connected with the BG – and to test whether the BG could be specifically implicated in language. Participants were required to rearrange materials sequentially based on linguistic (syntactic or conceptual) or non-linguistic (order switching) rules, or to repeat a previously ordered sequence as a control task. Functional magnetic resonance imaging (fMRI) data revealed a strongly active left-lateralized corticostriatal network, encompassing the anterior striatum, dorsolaterial and ventrolateral prefrontal cortex and presupplementary motor area, while the participants were sequencing materials using linguistic vs. non-linguistic rules. This functional network has an anatomical basis and is strikingly similar to the well-known associative loop implicated in sensorimotor sequence learning. We concluded that the anterior striatum has extended its original sequencing role and worked in concert with frontal cortical regions to subserve the function of linguistic sequencing in a domain-specific manner."

The following further findings are reported in this article and references therein:

The basal ganglia (BG) do not play a direct role in language control (according to the dominant view). Instead they may have more an assisting role – they contribute in cognitive control, enhancing selected activities while supressing competing ones, syntactic integration, language comprehension and switching between languages:

> "To date, the dominant view holds that the BG do not play a direct role in language mainly because damage to these areas alone does not consistently produce classical aphasic symptoms [...] and because the language deficits induced by BG damage can usually be traced to cortical hypoperfusion [...] Instead, the role of the BG may be in cognitive control, assisting language function by generally enhancing selected activities while suppressing competing ones [...] Indeed, monolingual studies have corroborated that the BG are involved in the controlled process of syntactic integration [...] and studies focusing on bilingualism have also shown that the BG are involved in second language comprehension and the control of switching between languages" (284)

The role of the corticostriatal loops is partially domain-general, but partially also specific to the domain of language (note: the striatum is a central part of the basal ganglia):

> "when subtracting out irrelevant common cognitive components and focusing on the linguistic sequencing process, we clearly found the BG involvement in a language-specific manner [...] a language-specific loop for sequencing is embedded in a domain-general sequencing (or more generally, cognitive control) system. The basic computation for cognitive control in general, or sequencing in particular, may be similar across domains, but in

processing linguistic materials, some language-sensitive component(s) may be added to the basic computation and thus a domain-specific process is created" (290)

The involvement of the basal ganglia in language control did most probably not develop as "new anatomical circuit de novo", but rather by "recruiting a preexisting BG loop", namely "the cognitive/associative/prefrontal loop":

"a corticostriatal loop, connecting the fronto-mesial structures to the head of the caudate nucleus, was important in the control (i.e. selection, inhibition and programming) of language and in modulating a large distributed network of language in the brain, including a ventral semantic stream, a dorsal phonological stream, a speech perception pathway, and an articulatory loop. Furthermore, the anatomy of this corticostriatal circuit is strikingly similar to that of one of the parallel BG loops in primates and humans – the cognitive/associative/prefrontal loop [...] This loop has been shown to be implicated in cognitive functions, such as sensorimotor sequence learning, attention, controlled retrieval and monitoring of information within working memory [...] Hence, our findings indicate that the linguistic sequencing process may be recruiting a preexisting BG loop, instead of creating a new anatomical circuit de novo, which is in accordance with the principle of preadaptation that old structures can assume a new function during evolution" (291)

Notes: term "fronto-mesial structures" refers to medial parts of the frontal cortex; terms striatal and striatum refer to a specific part of the basal ganglia; caudate nucleus and putamen are sub structures of the striatum.

Summarizing conclusions

The executive control functions of the brain and of the central nervous system can be attributed to the limbic system, basal ganglia, prefrontal cortex and anterior cingulate cortex.

That the limbic system represents emotions, moods, reward system, goals, motivations, and that it is also involved in metabolic homeostasis, senses of pain, neuroendocrine activity, autonomic functions and memory processes, is a matter of course, so that these facts must not additionally be substantiated here.

The concrete roles of the other parts of the executive control system are less obvious. But the previous descriptions and citations might have provided some insights into this matter, with the result that some interesting performance features can be made visible.

The basal ganglia supports among other things parallel control of different types of brain processes by anatomically segregated functional circuits, but which are also heavily interconnected. This comprises at least the following types of processes:

- skeletomotor control, in connection with the motor cortex
- oculomotor control, in connection with the frontal eye cortex,
- associative processes, in connection with the prefrontal cortex, and
- limbic processes, in connection with the anterior cingulate cortex.

The result is, that the brain is able to regulate these types of processes widely independently in parallel, but also with strong interdependencies.

This means among other things:

- Action can be done, controlled via the motor cortex

while associative/cognitive/memorizing processes are ongoing in parallel via the prefrontal cortex.

- Emotivational influences on both of these processes can be regulated to a great extent.

Another important outcome of the sections above refers to the evaluation and decision making processes which are ongoing in the brain. There is a decision making apparatus, which can be located in three cortical areas, the basal ganglia (ventral striatum), the medial frontal cortex (ventromedial prefrontal cortex and the medial orbitofrontal cortex), and in the anterior cingulate cortex. This system makes, firstly, decisions on which (reward) goal the attention should be spent, and secondly, which are the suitable actions to pursue this goal.

This system is able to make decisions very quickly, in terms of affective or impulsive decision making and also more slowly by evaluating the possible values, goals and actions thoroughly (value-guided decision making).

This system can make dual decisions – between two reward goals –, and also engage/search decisions – between an available reward and further search for another reward, but which may have a greater value, and it can involve several aspects into the evaluation process, such as reward probability, reward magnitude, average search values and effort costs.

An implicit implication, but which is not mentioned in the studies and in the sections above, should be that the original foundation of any value is provided by needs and that they are incorporated into the evaluation and decision process via the limbic system. Goals can only be caused by needs. Rewards are kinds of needs, which might not always be as imperative as basic needs, but which can nevertheless only take effect via the limbic system, the mesolimbic dopamine system and the caused emotions. A perceived object cannot become a goal if it not meets at least a small demand which is represented by a homeostatic, autonomic, painful, or dopaminergic stimulation in the limbic system and the basal ganglia.

It should be considered that the cortical structures, functions and processes, which are examined above, build also the system which provides humans with all kinds of cognitive, intellectual and artistic capabilities. It is the basis for all kinds of learning as well as for all types of decisions in daily life and about economic opportunities, political preferences and scientific hypotheses.

The examined articles make visible, that the decision making system is typically researched with animals or humans in simple situations with few rewards. But what about humans in real-life scenarios? It is all but impossible to investigate systematically what happens in the brain in complex or stressful real-life situations – with conflicts between a multitude of rewards and impairments –, and to detect interdependencies between brain processes and sociocultural developments. So this might be an open field, which requires further research.

Language is bound into the control processes via the associative control circuit. It is assumed that this control circuit, which is also available in animals, has been adapted in humans in a way, so that it additionally supports language control. This occurs partially in a domain-general way, but partially also specific to the domain of language.

Conclusions for the SPP-4DI model

The findings on the anatomically segregated parallel circuits in the basal ganglia (corticostriatal loops) and on the decision making apparatus are, together with the knowledge about the emotional-motivational system, highly consistent to the SPP model. They support the concepts *attention assessment controller* (AAC), *basal ganglia multiplexer* (BG multiplexer), *activity mode* and *preparation mode*. The concept *BG multiplexer* includes also the concepts *impulse control, intervention mechanism* and *inhibitive intervention*.

The reward goal selection part of the decision making process, which is executed in the basal ganglia and the medial frontal cortex, meets characteristic **attention** of the so called **attention assessment controller (AAC)** from the SPP model. Note: Characteristic *assessment* of the concept *attention assessment controller* fits more to the interplay between the emotional-motivational system and memory – more details see below in chapter "Emotion, motivation and memory".

The finding that motor control on the one hand and associative/cognitive/memorizing processes on the other hand can be controlled in parallel by segregated circuits in the basal ganglia, is highly consistent to what is called *activity mode* and *preparation mode* in the SPP model.

The **activity mode** refers clearly to real-time activity, "motor control" and the skeletomotor circuit. The **preparation mode** might comprise the associative processes (associative/cognitive control circuit), memory recall processes and the value-guided processes of the decision making apparatus. More or less directly, this is also related to declarative memory, secondary and tertiary associative cortical areas, cross-modal associations and cognitive processes. Procedural memory, which is actually to be attributed to the *activity mode*, is also present in the memory recall procedures, but only in the form of microrepresentations in the declarative memory, so typically in the region which is called hippocampus (see also chapter "Emotion, motivation and memory").

An important part of this system is language control. This comprises articulation, what can be attributed to the *activity mode*, if done with significant vigour. But the main part of the linguistic circuitry, which comprises also speech perception, semantics, abstraction and associations to other modes and modules of the cortex, is bound into the control processes via the associative/cognitive loop of the corticostriatal circuitry in terms of the *preparation mode*. It is likely that articulation plays also a role in this case, but if, then in the form of simulation, so largely non-vocal with inhibited vigour. That's why language simulation might be a typical ingredient of the *preparation mode* of the brain.

Term **basal ganglia multiplexer** (BG multiplexer; including the concepts *impulse control, intervention mechanism, inhibitive intervention* and *intervention switch*) refers to the interplay between the four anatomically and functionally distinct parallel basal ganglia circuits and to the dual decision making system, which combines affective/impulsive decision making via basal ganglia and value-guided decision making via frontal cortex and anterior cingulate cortex in integrated manner. *Impulse control* occurs, when an activity procedure, which is controlled by the skeletomotor circuit, is stopped in favour of further decision making via the cognitive/associative/prefrontal circuit. Motivational or emotional signals from the limbic basal ganglia circuit may play a role in the switch behaviour, just as results of perception and sensation processes.

The primary function of this multiplexer is to act as intervention switch for ongoing activity sequences that are getting inappropriate, but reinforcement, in terms of the opposite of intervention, completes the switch behaviour. The BG multiplexer works not only as on-off switch or as an intervene-reinforce switch, but rather as a modulation controller, which controls the vigour of each small piece of the activity sequences. That an activity sequence can be intervened totally and very quickly (before a first external sign is produced) is only one of the capabilities of this brain feature.

But the finding, which is actual important in relation to the SPP model, is, that the *basal ganglia multiplexer* supports the interplay between *activity mode* and *preparation mode* of the brain processes. Both types of processes occur constantly in parallel, controlled and synchronized by the basal ganglia circuits.

Consideration on the social reference of impulse control

The capability to avoid inappropriate behaviours and to search for more suitable or more convenient behaviours might primarily have been forced in social contexts, and the impulse control function, the BG multiplexer and the distinct control mechanisms of the human brain processor might primarily have been developed for regardfulness towards the social environment rather than the natural environment. This is only consistent, when considering that survival of the group in a hostile environment was the actual challenge in the history. Inconsiderate exploitation of natural resources was typically no problem as long as this was connected with benefits for the group. This is as it were inscribed into our neural apparatus that nature is definitely able to defend its own interests, so that it can be affected with no regret.

But this might slightly turn into the opposite due to our massive impairment of biodiversity, so that we have to learn how to perform impulse control also towards nature.

Reminder: This means "that inappropriate responses can be stopped very quickly and perfectly [...] a special 'hyperdirect pathway' has been developed for this purpose by the Evolution" (see above under "The motor circuit").

Emotion, motivation and memory

Note: This chapter summarises basic knowledge from various popular sources (without known references) and from Popper and Eccles 2006 and Gazzanega and Mangun 2014.

Emotion and Motivation

Preliminary, it is difficult to find a comprehensive description about the neural mechanisms of emotion and motivation in scientific papers. The so called limbic system, which was proposed for this topic in the past, is no longer referred much by contemporary scientific views. But it can nevertheless give an orientation about which areas and circuits are involved in this part of the control processes.

There are four main systems which are involved in emotional and motivational processes, but which are highly integrated. These are the autonomic interface, involving homeostatic regulation, the sensory system (particularly pain sensation), emotional processing, and the reward system.

1) The **autonomic interface** integrates the nervous system with the autonomic nervous system and the metabolic processes of the body. The main interfaces for these processes are a subcortical area which is called hypothalamus and the endocrine system, including a larger collection of glands, like pituitary gland, pineal gland, thyroid gland, adrenal gland, pancreas gland, ovaries gland, testes gland and others.

The core part of the control processes which occur on the behalf of these interfaces is called **homeostasis**. This comprises, according to Maslow 1987, 15, and references therein, the regulation of metabolic balances like "(1) the water content of the blood, (2) salt content, (3) sugar content, (4) protein content, (5) fat content, (6) calcium content, (7) oxygen content, (8) constant hydrogen-ion level (acid-base balance), and (9) constant temperature of the blood [...] minerals, the hormones, vitamins, and so on". This is linked to feelings like hunger, thirst, cold, hot, sexual arousal etc. This includes also a control loop that is called hypothalamic-pituitary-adrenal (HPA) axis and which is connected with the regulation of stress levels in the nervous system and the metabolic system in consistent manner.

This includes also regulation of the balance between sympathetic and parasympathetic nervous system, which are two parts of the autonomic nervous system. The former represents preparedness for action and the latter rest and digestion, and the autonomic regulation processes include always regulation of the balance between both aspects of life, and thus between sympathetic and parasympathetic nervous system.

2) The **sensory system** has two main aspects:
- Information: provide signals for sensorimotor control or for the informational aspect of cognitive processes and decision making processes (see something, smell something or feel something which is not painful).
- Motivation: **pain sensation**.

Only the latter aspect is of interest here. This refers primarily to the sense of touch and the somatosensory system, whose signals are typically bound into cortical processes via a subcortical area which is called thalamus. This comprises sensory signals representing, for example, touch, vibration, heat, (the somatosensory component of) balance and pain. However, only pain is of interest here. This takes partially effect via explicit pain receptors in body and skin, and partially via the normal receptors if they are confronted with overmodulation, e.g. due to too much pressure or heat.

Besides this there are signals of pain which are produced by other senses, like vision (dazzling light), hearing (loud noise) or taste (bitter). But these are difficult cases – the transition between information and pain is widely adaptable, and domain specific investigation is needed for understand how these types of perception patterns turn into pain and motivational stimuli.

3) **Emotional processing** is particularly attributed to a brain area which is called amygdala. This is associated with extraordinary excitation states that are caused by alarming environmental stimuli. This comprises fear and anxiety, but also pleasure and delight and a wide range of other feelings, which occur due to interactions with the natural and social environment. The amygdala, and thus emotional processing, is closely linked to other areas and systems, like habenula, which represents the olfactory sense, the autonomic interface or the reward system (see next paragraph).

4) The **reward system** is connected with concepts like mesolimbic dopamine system or mesocorticolimbic dopamine system. It represents emotions and the pursuit of any type of good feelings or rewards, at which this is seen as an open system with no determination or restriction, what kind of desires might occur with this behalf. The reward system is mainly represented by the following subcortical components: amygdala (representing emotions), septal nuclei, nucleus accumbens, ventral tegmental area and pathways to several substructures of the basal ganglia (striatum, caudate nucleus, putamen, substantia nigra etc.). Neurotransmitters, which play an important role in this system, are dopamine, serotonin, glutamate and gamma-aminobutyric acid (GABA).

These four parts of the emotional-motivational system, mainly represented by the subcortical areas thalamus, habenula, hypothalamus, amygdala, septal nuclei, nucleus accumbens, ventral tegmental area and basal ganglia (as described above) are interconnected with several cortical areas, that are linked to memory, behaviour control and cognitive control. These are mainly the regions which are called entorhinal cortex, hippocampus, cingulate cortex and prefrontal cortex. Most of the pathways and loops of this system are cross-linked via the structures of the basal ganglia. Thus, the basal ganglia is the crucial key component of the emotional-motivational system of the brain. Each type of emotion, motivation and need is conveyed via this intermediary device.

The entorhinal cortex and the hippocampus are key areas of the memory system, particularly of the declarative part. The entorhinal cortex is seen as hub to all other areas of the memory system. The hippocampus represents episodic memory and spatial memory. More details see in section "The types of memory".

The cingulate cortex and the prefrontal cortex are, in conjunction with the basal ganglia, the crucial parts of the decision making system. More details see in section "Decision making in BG and frontal cortex".

The types of memory

The brain, particularly the cortex, is first of all a great memory system, which adapts continuously to the requirements of daily life. This is true for the plasticity of basic or subcortical structures, cascades of modules performing perception and processing of sensory information, control systems as well as for the areas and functions that are explicitly attributed to kinds of memory. Concentrating on the latter, one can start with the concepts of procedural and declarative memory.

Procedural memory happens in the sensory cortex, the motor cortex, adjacent associative areas and further intermediary parts of the brain. This includes all nerve cords and modules that are needed for the execution of activity sequences. The cerebellum as great supporter of motor coordination and smooth movements should also be mentioned here.

This type of memory is not, like declarative memory (see below), directly accessible by procedures of remembering or conscious thinking and recombination. It can only be recalled, modified and trained by activity. It is present in the declarative memory by microrepresentations that can be used as queues for trigger the appropriate procedural activity sequences when required (see below).

In contrast, **declarative memory** is the part which can explicitly be recalled and involved into all kinds of conscious thought processes, recombination processes and associative processes. It is mainly located in the so called temporal lobe, the hippocampus, the entorhinal cortex, in secondary, tertiary and unspecific associative areas as well as in the language areas, which can also be characterised as associative to a great extent.

The associative areas are the actual locations of (declarative) **long term memory**. Important key mechanisms for this part of the memory system are "Cross-modal associations" and "Integrative encoding …" – more details see in the appropriate sections below.

The key areas of the declarative memory are the hippocampus and the entorhinal cortex. The **hippocampus** stands for spatial memory (navigation), episodic memory and the intermediation between short term memory and long term memory (see also the sections about "Integrative encoding …"). It is known for the ability to keep episodic information for a couple of weeks. After this period, the information is either transferred to the long term memory via repetitive excitation and integration into the system of enduring associative memories, or it is forgotten (more details see below under "Integrative encoding of memories over time"). The hippocampus is also known for extraordinary high plasticity – growth of nerve cells and synapses can occur in this region with high intensity into old age.

The **entorhinal cortex**, which is also called limbic association cortex and which is located in the medial temporal lobe, is another significant area of the memory system. It is associated with the following characteristics:

- It is the main interface between hippocampus and neocortex.
- It is responsible for memory navigation.
- It functions, respectively, as multimodal association area or as a hub in a widespread network of memory.
- It is a multimodal association area, supporting unspecific associations between various modules and association areas with more domain specific determination.
- It supports the perception of time.

These key areas of the declarative memory system, so hippocampus and entorhinal cortex, are also the primary gateways for incorporate value information from the emotional-motivational system into memorised information. More details see in sections "Integrative encoding of procedural, emotional-motivational, sensory-perceptual and internal thought components" (below) and "Emotion and Motivation" (above).

Note: Further areas, as e.g. the perirhinal and parahippocampal cortices may also be involved in the declarative and episodic part of the memory system.

Short term memory does not exist in terms of a static memory store with clear location. This concept refers rather to the attention-related processes that are constantly ongoing in prefrontal-basal-ganglia-cortical loops and namely in the decision making system (see also Ranganath, Hasselmo and Stern 2014 and Glimcher 2014). These processes circle always around few topics that are currently attracting attention due to preceding attentional and evaluative processes, to external stimuli, elicited via the perception system and due to prevailing needs, elicited via the emotional-motivational system (see also the descriptions in chapter "Basal ganglia (BG) and frontal cortex" and in section "Emotion and Motivation"). This includes the immediate control of current activities and also all processes of value-guided decision making, association and cognition. The prefrontal-basal-ganglia-cortical circuits, including the decision making circuits, are able to switch the attention short term between several tasks, what is partially (or traditionally) linked to the concept *short term memory*. The ability to re-encourage former excitation patterns directly in this system reaches to durations under one minute while recall after longer passivity is also possible, but only via processes of episodic memory or (long term) associative memory.

The further descriptions in this chapter refer primarily to the processes and encodings in the declarative part of the memory system.

Cross-modal associations

The specific role of associative areas for memory and cognition has already been emphasised in the classic publication from Popper and Eccles 2006 in "Chapter E2 Conscious Perception", section "9. Cutaneous Perception (Somaesthesis)", subsection "9.3. Secondary and Tertiary Sensory Areas", 260: "In palpation there is first the shaping of the hand for grasping an object, and secondly the moving of the hand over the surface of the object in an active exploration. In this way cutaneous sensing leads to feature detection that matches the visual feature detection in the inferotemporal lobe". Incorporating language into the system of interconnections has two aspects – it adds another ability of the brain to develop interconnection complexity and it opens the social dimension. Popper and Eccles 2006 write in "Chapter E4 The Language Centres of the Human Brain", section "24 Résumé", 295f.: "Particular importance is attached to Brodmann's areas 39 and 40, which came very late in evolution, being barely recognizable in nonhuman primates. These are the areas specifically concerned in cross-modal associations, that is associations from one sensory input, say touch, to another, say vision [...]. It is postulated that language comes when you have the association between objects that you feel and objects that you see, and which you then name. [...]. Language provides the means of representing objects abstractly and for manipulating them hypothetically in one's mind."

By this way, cross modal associations enable the brain to build an internal representation of the world via perception, active exploration, storing the resulting experiences in sensory and motor areas of the cortex and by integrating these primary contents in associative areas. This includes listening, reading, speech, writing and appropriate language areas. The result is that the structure of the world can be internalized to a great extent by internal synaptic representations, that this world view can be exploited for make attitudes and behaviours highly suitable and that this can be combined with social interconnectedness.

Integrative encoding of memories over time

The capability to exploit cross-modal associations gets another dimension, when considering the temporal aspect and the fact that experiences and associations, once made, are complemented again and again by further experiences and associations. This dimension of neural processes comprises that new experiences are integrated into the network of existing memories, that existing memories are strengthened again by this way and that novel inferences are produced. The cortical region, which is called hippocampus, plays a crucial role in these processes. This is described by Davachi and Preston 2014 and references therein under headline "Memory reinstatement during new encoding" (543f.):

"As highlighted in the previous section, memories for past events are often reinstated during new experiences. These findings emphasize that memory encoding and retrieval are not performed in isolation, but rather are interactive processes. Recent human neuroimaging research indicates that new learning influences, and is influenced by, reactivation of existing memories. In one such study, participants learned overlapping (e.g., watch – sink and later watch – pipe) and nonoverlapping (e.g., peanut – moose) pairs of pictures […] The findings revealed that the degree to which prior memories (e.g., memory for the watch – sink event) were reinstated during encoding of new overlapping experiences (watch – pipe) was associated with greater retention of the originally learned information when compared to memory for nonoverlapping information. Moreover, hippocampal engagement during encoding of the overlapping pairs was related to both cortical memory reinstatement and improved memory retention. These findings indicate that memory reinstatement during new encoding helps reduce forgetting of past events. One possibility is that reduced forgetting was not just the result of strengthening of reactivated memories, but may also have resulted from a hippocampal-mediated integrative encoding mechanism, whereby newly encountered information is integrated with existing memories at the time of learning [...].

According to this mechanism, the fundamental role of the hippocampus in memory is not only to form relationships among elements within an individual experience, but also to construct memory representations that link memory elements across discrete experiences. [...] Integrative encoding proposes that when a new event shares a common feature, or features, with an existing memory trace, the common features elicit memory reinstatement through hippocampal pattern completion. New experiences would then not only be encoded in the context of presently available information in the external world, but would also be bound to reactivated representations of prior related memories. Furthermore, this mechanism makes the fundamental prediction that by combining information across discrete events, hippocampal memory representations would include information that goes beyond direct experience. Accordingly, integrative encoding would not only support strengthening of existing memories [...], but would also support novel inferences about the relationships between memory elements that were experienced at different times and generalization of knowledge to entirely new situations, a hallmark of episodic memory."

Integrative encoding of procedural, emotional-motivational, sensory-perceptual and internal thought components

The hippocampus is particularly also known as supporter of connections to the **procedural memory**. Episodic memory means mainly, that perception events and sensorimotor activity sequences are recorded and incorporated into a timeline. In addition, it is possible to reinstate this information, include it into associative thought processes and exploit it for any further purpose. This comprises also that it must be possible to include procedural memory content in any form into these processes and to reactivate the appropriate activity sequences at any time again – in terms of planning, usage/execution, training/optimisation or adaptation. This requirement is known as being implemented by microrepresentations of memorized procedural sequences that can be included into the recordings in the episodic memory. These microrepresentations function as keys for the ability to reactivate the referred procedural activity patterns at any time again (see also below under "Microrepresentations …").

Another point is that also **emotional-motivational value information** is always integrated as coherent ingredient into the processes and encodings in the declarative memory. This occurs via the pathways between the emotional-motivational system (details see in section "Emotion and Motivation") and the key areas of the declarative part of the memory system, which are the hippocampus and the entorhinal cortex (see also above under "The types of memory" and below under "Microrepresentations …").

Sensory-perceptual experiences and **internal thoughts** occur, like activity sequences, primarily in appropriate domain specific cortical areas and they are involved into episodic recordings and further memory processes via microrepresentations (see also below).

Microrepresentations of sensory-perceptual experiences, actions, internal thoughts, and emotions in the episodic memory

Davachi and Preston 2014, focusing on episodic memory in the medial temporal lobe (MTL), report about a "memory as reinstatement model" (MAR model), which is proposed as "a summary of the common elements of many existing models" (Davachi and Preston 2014 and references therein, 539f.):

"During an experience, the MAR model proposes that the ongoing, dynamic representation of that experience is represented in distributed cortical and subcortical patterns of neural activation. These activation patterns are driven by sensory-perceptual (visual, auditory, somatosensory) experience, actions, internal thoughts, and emotions, to name a few. Thus, neural activation patterns at any one time point can be thought of as representing the current episode and state of the organism. Critically, it is thought that the distributed pattern of cortical and subcortical firing filters into the MTL cortices and converges on the hippocampus, where a 'microrepresentation' of the current episode is created. [...] What results from a successfully encoded experience is a hippocampal neural pattern (HNP) and a corresponding cortical neural pattern (CNP). Importantly, the HNP is thought to contain the critical connections between representations that allow the CNP to be accessed later and attributed to a particular event.

Importantly, episodic memory retrieval is thought to

involve cue processing that, if successful, will lead to a re-instatement of the HNP. Retrieval cues (e.g., an external stimulus or internal thought) are thought to serve as 'keys' that unlock the HNP associated with a prior experience, a process referred to as hippocampal pattern completion. Pattern completion refers to the idea that a complete pattern (a memory) can be reconstructed from only a subset of the elements making up that pattern. Thus, successful retrieval is thought to involve reinstatement of all or some of the HNP established during encoding. Finally, reinstatement of the HNP is then thought to be instrumental in reinstating the corresponding CNP, resulting in the concurrent reactivation of disparate cortical regions that were initially active during the experience. Importantly, it is thought that this final stage of cortical reinstatement underlies the subjective experience of recollection and drives mnemonic decision making."

Conclusions for the SPP-4DI model

The four parts of the emotional-motivational system, which are described as autonomic interface, sensory system (particularly pain sensation), emotional processing and reward system (see section "Emotion and Motivation"), are the providers of purpose for the brain. They ensure that the brain works at the service of the creature, in which it is implemented. This occurs partially in terms of strict forces (homeostasis, pain) and partially in terms of pleasure and joy with large creative leeway (rescue from suffering, positive emotions, reward system). But both types of mechanisms take effect in closely interconnected manner via a joint system of crosslinks in the basal ganglia. This is consistent to the concepts *attention assessment controller* (AAC) and *basal ganglia multiplexer* (BG multiplexer; including the concepts *impulse control, inter-*

vention mechanism and *inhibitive intervention*) which are proposed in the SPP model.

It can be assumed that the assessment part of the attention assessment controller is bound into the memory system via the pathways from the emotional-motivational system to the memory system. Memory means storing information on the basis of evaluation or rating – good results in excitation and bad results in inhibition – what is consistent to the term assessment.

It can further be assumed that the attention part, what is primarily meant in the sense of selective attention or attention to (German "Zuwendung"), occurs via the decision making system.

It is further proposed, that the assessment ratings are always established as an inherent value part of memory – as degrees of positive or negative emotional-motivational experiences, encoded in the excitation effects and inhibition effects of the involved synapses – and that these values are exploited by the decision making between concurrent candidate patterns in relation to pursued values.

This is consistent to the theses
- that the brain does not store any information without purpose and value and
- that it does not select any stored information without purpose and value.

That purposes and values can be multifarious – reaching from pure pleasure in learning to the fight against misery and distress – is another story.

Cognitive control and emotions

Seesaw models and more advanced approaches

Ochsner (2014) reports about seesaw models of cognitive control and emotion (see in Ochsner 2014 and references therein). He starts with descriptions of the basic idea behind these seesaw models, which emphasise the contrariness between cognitive control and emotion:

"Historically, the idea that cognition and emotion seesaw back-and-forth with one another dated back at least to Descartes's ideas about the dichotomy between passion and reason". (719)

"Both, human and animal research have reliably associated cognitive control and emotional responding with distinct but overlapping sets of brain systems." (720)

But in the end he comes to different conclusions with insights into interdependencies between both components under the following headline (726):

"Beyond the seesaw: An alternative account of the relationship between cognitive control and emotion, and its implications for theory and research

This chapter began with the compelling view that emotion and cognitive control are often in opposition. As a shorthand, we referred to these theoretical accounts as seesaw models that posited the engagement of cognitive control in the pursuit of explicit or implicit regulatory goals as the effect of diminishing neural and behavioral signatures of affective response.

Using the glossy, popular version of these models as a foil, we then reviewed four kinds of evidence that cognitive control in affective processes can interact in ways not well accounted for by seesaw models [...]. First, we saw that the use of cognitive control to elaborate the affective meaning of an event can generate emotion via the recruitment of prefrontal control systems and the triggering of activation in subcortical affect systems like the amygdala. Second, we saw that control regions could help select stimulus-appropriate words for describing and reporting on one's current emotional state. Third, in regard to perceiving other people's emotions, we saw that control regions may help resolve social cognitive conflicts between interpretations of different kinds of cues that provide different kinds of information about others' emotions. Fourth and last, we saw that prefrontal control regions can be used to up-regulate emotion, and in so doing, up-regulate responses in subcortical affect systems like the amygdala.

If these data either are inconsistent with or fall outside the intended explanatory realm of seesaw models, then how should we conceive of cognitive control – emotion interactions?"

"CONSTRAINT SATISFACTION AND THE MAKING OF MEANING In their classic book, *The Measurement of Meaning*, Osgood, Suci, and Tannenbaum showed that approximately two-thirds of the meaning we ascribe to things in the world arises from our valenced evaluation of those things as good or bad, positive or negative, pleasant or unpleasant (Osgood, Suci, & Tannenbaum, 1957). Fundamentally, this highlights that our affective responses form a core element of what events, people, places, things, and so on mean to us. Indeed, the most profound and important experiences of our daily lives wouldn't mean as much if they were bereft of their emotional potency."

In the following, Ochsner (2014) and references therein propose, as an alternative, a "constraint satisfaction model", which is able to deal with the "cognitive control – emotion interactions" as a feature.

In the end, Ochsner (2014) comes to the following conclusions among other things (728):

> "WHAT'S NEXT? While this model relies as much on metaphor as a seesaw model, it may provide a framework for accounting for a broader range of interactions between cognitive control and emotion. As such, it suggests two directions for future research.
>
> First, we might expand our experimental purview to study not just cases where cognitive control appears to down-regulate some type of affective (or more generally, maladaptive) response, but to study the myriad other situations where we generate, introspect on, perceive, and regulate emotion in ourselves and others in ways that enhance, transform, integrate, and arbitrate between potential affective responses, interpretations, judgments, and so on. Second, it highlights just how far future research needs to go in providing increasingly specific descriptions of underlying neural mechanisms. As a field, we are increasingly moving beyond descriptions of neural bases that specify only relative levels of activation in different systems, a la seesaw models. Instead, more complicated path, mediation, and multivoxel pattern-based models are appearing that specify complex combinations of systems and the way in which they interact."

Finally, Ochsner (2014) characterises the aims for future research as follows:

> "understand the neural underpinnings of self-regulatory success and failure" and "go where no researchers have gone before, exploring new ways in which cognitive control and emotion can interact." (728)

Conclusions for the SPP-4DI model

The 4DI model, which emerges from the SPP model as a consistent result, follows the intentions of Ochsner (2014), but with the peculiarity that this takes place on a very abstract level. The 4DI model is based on the idea that reason, cognition and pure informational intelligence on the one hand and emotion, vigour and passion on the other hand is indeed in a contradictory relationship, but that real intelligence (of the 4DI type) is provided by the ability to combine both components in successful manner. This is seen as the crucial tension field for the development of human intelligence.

The SPP-4DI model, on its abstract level, can only develop as a link to other disciplines, like psychology, social sciences, and so on, but it will not be able to enlighten neurophysiological details – this task can rather only be addressed by approaches like the constraint satisfaction model or other efforts in neuroscientific research.

It should be mentioned that the models SPP and 4DI are based on a concept of emotion (or "emotivation") that might be slightly different to the one that is adopted in Ochsner 2014 and in other articles in Gazzanega and Mangun 2014.

Consciousness

Block, Ned: Comparing the Major Theories of Consciousness

Ned Block (2009) compares three theories of consciousness:
* The higher order state approach, or higher order thought (HOT) approach.
* The global workspace account of consciousness.
* The biological theory.

Block provides a discussion on how these theories try to explain consciousness and what are the remaining discrepancies and open questions. The "explanatory gap" between phenomenal consciousness and its physical basis is made a central topic and incorporated as a crucial benchmark criterion into the discussion.

How the explanatory gap is described in Block 2009 and how it is seen from the perspective of the SPP model is already covered in the section "Conscious experiences" in Part 1.

Regarding the three above-mentioned theories, Block comes to the following results (among other things):
* According to the higher order state approach, (Block 2009, 1114):

"The fact that the HOT theory cannot recognize the real explanatory gap makes it attractive to people who do not agree that there is an explanatory gap in the first place – the HOT theory is a kind of "no consciousness" theory of consciousness. But for those who accept an explanatory gap (at least for our current state of neuroscientific knowledge), the fact that the HOT theory does not recognize one is a reason to reject the HOT theory. The HOT theory is geared to the cognitive and representational aspect of consciousness, but if those aspects are not the whole story, the HOT theory will never be adequate to consciousness."

- According to the global workspace account of consciousness, (Block 2009, 1114):

  "This very short argument against the HOT approach also applies to the global workspace theory, albeit in a slightly different form. According to the global workspace account, the answer to the question of why the neural basis of my experience of red is the neural basis of a conscious experience is simply that it is globally broadcast. But why is a globally broadcast representation conscious? This is indeed a puzzle for the global workspace theory but it is not the explanatory gap because it presupposes the global workspace theory itself, whereas the explanatory gap (discussed previously) does not."

- According to the biological theory, (Block 2009, 1114):

  "The biological account, by contrast, fits the explanatory gap – indeed, I phrased the explanatory gap in terms of the biological account, asking how we can possibly understand how consciousness could be a biological property. So the biological account is the only one of the three major theories to fully acknowledge the explanatory gap."

How the biological theory may explain consciousness is proposed by the example of area MT+ in the visual cortex in Block 2009 and references therein, 1112f.:

  "Visual area MT+ reacts to motion in the world, different cells reacting to different directions. Damage to MT+ can cause loss of the capacity to experience this kind of motion; MT+ is activated by the motion aftereffect; transcranial magnetic stimulation of MT+

disrupts these afterimages and also can cause motion 'phosphenes' [...]. However, it is important to distinguish between two kinds of MT+ activations, which I will call nonrepresentational activations and representational activations. [...] However, if activations of MT+ are strong enough to be harnessed in subjects' choices (at a minimum in priming), then we have genuine representations. [...]

What makes such a representational content phenomenally conscious? One suggestion is that active connections between cortical activations and the top of the brain stem constitute what Alkire, Haier, and Fallon (2000) call a 'thalamic switch.' There are two important sources of evidence for this view. One is that the common feature of many if not all anesthetics appears to be that they disable these connections (Alkire & Miller, 2005). Another is that the transition from the vegetative state to the minimally conscious state (Laureys, 2005) involves these connections. However, there is some evidence that the 'thalamic switch' is an on switch rather than an off switch (Alkire & Miller, 2005) and that corticothalamic connections are disabled as a result of the large overall decrease in cortical metabolism (Velly et al., 2007; Alkire, 2008; Tononi & Koch, 2008) – which itself may be caused in part by the deactivation of other subcortical structures (Schneider & Kochs, 2007). Although this area of study is in flux, the important philosophical point is the three-way distinction between (1) a nonrepresentational activation of MT+, (2) an activation of MT+ that is a genuine visual representation of motion, and (3) an activation of MT+ that is a key part of a phenomenally conscious representation of motion."

So it can be concluded that the biological theory may be

able to provide useful solution approaches on the matter of consciousness and the explanatory gap. Deeper analysis in this direction would certainly be recommended. This refers to further empirical research, but also to the formation and verification of hypotheses and models. Whether there may be a correlation between the biological theory approaches and the SPP model still has to be investigated.

The integrated information theory of intelligence/consciousness (for more details see the next section) is classified by Block as a variation of the global workspace account of consciousness, sharing the gaps and discrepancies to a certain extent.

Tononi, G.: The integrated information theory of consciousness

The integrated information theory of consciousness (IIT, Tononi 2012) tries to clarify how information is processed by the brain to obtain conscious experiences from it. It is based on certain axioms and postulates. "The information axiom asserts that every experience is specific - it is what it is by differing in its particular way from a large repertoire of alternatives. [...] The integration postulate states that only information that is irreducible matters [...] The exclusion postulate states that only maxima of integrated information matter" (Tononi 2012, 290, Abstract). An important basic statement is: "Information can best be defined as a 'difference that makes a difference'" (Tononi 2012, 294, referring to Bateson 1972). What IIT tries to achieve is described in the section "Conclusion": "IIT attempts to provide a principled approach for translating the seemingly ineffable qualitative properties of phenomenology into the language of mathematics" (Tononi 2012, 318).

IIT explains "Consciousness as integrated information" (Tononi 2012, 290). This means that a system which is able to integrate information – in terms of maxima of integrated information – to a certain extent and which is able to differ this information from other integrated information, is capable to produce phenomenal consciousness. This refers to any piece of matter or to any information processing engine with a minimal level of complexity, independent of whether this system is a living organism or whether it is equipped with any type of self-regulation or not.

Tononi focuses on pure information processing capabilities in terms of awareness, cognition or informational intelligence. So IIT considers only the informational aspect of the brain processes. It does not ask how aspects like behaviour, autonomic regulation, motivation, pain, emotion, empathy or passion are incorporated into the equation, what would certainly be necessary to complete the picture.

IIT emphasises that consciousness depends on the fact that information is integrated and that experiences are made which are different to alternatives. But the question is not asked of why the integration of information, obtained from different senses, and the comparison of experiences to alternatives, could be relevant at all. But there should be a reason *why* information is processed as explained by IIT. So consciousness will actually require an evaluative component. That information makes a difference is only possible if there is a reason for making the difference, if there is any purpose behind it or any kind of (self-) regulation. Information is just chaos which does not matter if there is no reason and no relevance, independent of whether this could be integrated or not, and if there might be alternatives or not.

That information makes a difference might correlate with the concepts of homeostasis, basic and artificial needs and emotivational evaluation as explained by the SPP model.

Evaluative information which is provoked by the autonomic nervous system, the sensation of pain or the mesolimbic dopamine system and incorporated into neural excitation patterns, could also be regarded as information. So one could assume that the evaluative parts of the information are seen as making the decisive difference between conscious and not conscious. But the IIT theory is either not proposed to be interpreted in this way, or the specific role of evaluative information is seen as a basic precondition, but one that is not explicitly mentioned in Tononi 2012.

But it should also be clear that the fact, that a (biological) system is equipped with self-regulation and purpose, does not implicitly make this system conscious. It is an important lesson, not least from IIT, that self-regulation must be accompanied by a certain degree of informational complexity, in the sense of integration and differentiation, to achieve conscious states.

Popper/Eccles: The Self and Its Brain

Popper and Eccles (2006) provide a major comprehensive description of the nervous system and of the brain from the philosophical viewpoint (Popper) and from the neurophysiological and neurological viewpoint (Eccles). The book was first published in 1977, but it represents a great part of the state of knowledge, which can today still be regarded as valid.

A good explanation of what the conscious mind is is provided in particular in the Eccles part of the book.

His model is built around the concept of the "liaison areas", recognisable by the following statement:

"The self-conscious mind is actively engaged in reading out from the multitude of active centres at the highest

level of brain activity, namely the liaison areas of the
dominant cerebral hemisphere. The self-conscious mind
selects from these centres according to attention, and
from moment to moment integrates its selection to give
unity even to the most transient experiences." (Popper/
Eccles 2006, 362)

But this becomes difficult in connection with the following
statements in the Eccles section, which are questionable:

"But the question: where is the self-conscious mind
located? is unanswerable in principle. This can be ap-
preciated when we consider some components of the
self-conscious mind. It makes no sense to ask where are
located the feelings of love or hate, or of joy or fear,
or of such values as truth, goodness and beauty which
apply to mental appraisals." (Popper/Eccles 2006, 376)

The SPP model provides a different answer to this question:
The areas where experiences and feelings occur can be
located – they are largely superimposable to the areas
where perception is integrated with associative/cogni-
tive processes and emotivational evaluation in the preparation mode of the attention assessment controller.

Chalmers, David J.: The Character of Consciousness

(1)

Chalmers (2010) explains in the introduction:

"Consciousness is an extraordinary and multifaceted
phenomenon whose character can be approached from
many different directions. It has a phenomenological

and a neurobiological character. It has a metaphysical and an epistemological character. It has a perceptual and a cognitive character. It has a unified and a differentiated character. And it has many further sorts of character.

We will not understand consciousness by studying its character on just one of these dimensions." (Chalmers 2010, xi)

So Chalmers makes consciousness a major complicated issue. But in fact it is not. According to the SPP model "consciousness is" simply "information enlightened by emotivational evaluation", based on achievements (a), (b) and (c), as explained in the section "Conscious experiences" in Part 1.

Chalmers's examination of the issue from different directions offers a significant discussion of the philosophical aspects of the question of consciousness. But they are not necessarily needed to explain what consciousness is and why it occurs, at least not from the analytical perspective of the SPP model, which is preferentially based on neurobiological findings.

(2)

Important questions arising in relation to two different perspectives are discussed in Part II "The science of consciousness", Chapter 2 "How can we construct a science of consciousness?" (Chalmers 2010, 37ff.):

"The task of a science of consciousness, as I see it, is to systematically integrate two key classes of data into a scientific framework: *third-person data*, or data about behavior and brain processes, and *first-person data*, or data about subjective experience." (Chalmers 2010, 37)

"Third-person data concern the behavior and the brain processes of conscious systems. These behavioral and neurophysiological data provide the traditional material of interest for cognitive psychology and cognitive neuroscience. Where the science of consciousness is concerned, some particularly relevant third-person data are those concerning perceptual discrimination and those involving verbal reports. Direct measurements of brain processes also play a crucial role in cognitive neuroscience, of course.

First-person data concern the subjective experiences of conscious systems. It is a datum for each of us that such experiences exist: we can gather information about them both by attending to our own experiences and by monitoring subjective verbal reports about the experiences of others. These phenomenological data provide the distinctive subject for the science of consciousness. Some central sorts of first-person data include those having to do with the following:

- visual experiences (e.g., the experience of color and depth)
- other perceptual experiences (e.g., auditory and tactile experience)
- bodily experiences (e.g., pain and hunger)
- mental imagery (e.g., recalled visual images)
- emotional experience (e.g., happiness and anger)
- occurrent thought (e.g., the experience of reflecting and deciding)

Both third-person data and first-person data need explanation." (Chalmers 2010, 38)

"The problems of explaining third-person data associated with consciousness are among the 'easy' problems of consciousness [...]. The problems of explaining first-person data is the hard problem." (Chalmers 2010, 38f.)

"The lesson is that *as data*, first-person data are irreducible to third-person data and vice versa. That is, third-person data alone provide an incomplete catalogue of the data that need explaining: if we explain only third-person data, we have not explained everything. Likewise, first-person data alone are incomplete. A satisfactory science of consciousness must admit both sorts of data and must build an explanatory connection between them." (Chalmers 2010, 39)

How can the SPP model be related to these two perspectives?

The SPP model tries to explain how the decisive performance features of the nervous system and of the brain are achieved. This is primarily done from a distant perspective which might correlate with third-person data according to the citations above. But the SPP model also tries to explain how first-person data like experiences, thinking and emotions can evolve and how they are integrated. This can be seen as an explanatory connection between both sorts of data. What is not examined are the concrete phenomena in the world of first-person data.

But it is clear that these phenomena cannot be seen as reducible to some functional principles of the suitability probability processor (SPP). The SPP can rather be seen as a medium for the projection of phenomena from the external environment and from the body into appropriate parts of the neural networks of the brain. That the resulting associative-emotivational patterns can also be combined with any form of hypothetical projection, which is completely decoupled from real interaction contexts and which can even be fictional or far away in space or time, has the consequence that there is no limitation for creative potential in relation to first-person data and appropriate experiences (see also sections "Needs and suitability probability evaluation" and

"Conscious experiences" in chapter "The SPP model" in Part 1).

(3)

What is seen as the key question and the really hard problem of consciousness is explained in Chalmers 2010 in Part I "The problems of consciousness", Chapter 1 "Facing up to the problem of consciousness" (5ff.):

"The really hard problem of consciousness is the problem of *experience*.

When we think and perceive, there is a whir of information processing, but there is also a subjective aspect. As Nagel (1974) has put it, there is *something it is like* to be a conscious organism. This subjective aspect is experience. When we see, for example, we *experience* visual sensations: the felt quality of redness, the experience of dark and light, the quality of depth in a visual field. Other experiences go along with perception in different modalities: the sound of a clarinet, the smell of mothballs. Then there are bodily sensations from pains to orgasms; mental images that are conjured up internally; the felt quality of emotion; and the experience of a stream of conscious thought. What unites all of these states is that there is something it is like to be in them. All of them are states of experience.

It is undeniable that some organisms are subjects of experience, but the question of why it is that these systems are subjects of experience is perplexing. Why is it that when our cognitive systems engage in visual and auditory information processing, we have visual or auditory experience: the quality of deep blue, the sensation of middle C? How can we explain why there is something it is like to entertain a mental image or to experience an emotion? It is widely agreed that experience arises from a physical basis, but we have no good explanation of why and how it so arises. Why should physical processing give rise to a rich inner life at all? It seems objectively unreasonable that it should, and yet it does.

If any problem qualifies as *the* problem of consciousness, it is this one." (Chalmers 2010, 5)

Chalmers continues by discussing some attempts to explain this key problem, leading, among other things, to the following conclusions:

"To explain experience, we need a new approach. The usual explanatory methods of cognitive science and neuroscience do not suffice. These methods have been developed precisely to explain the performance of cognitive functions, and they do a good job of it. Still, as these methods stand, they are equipped to explain *only* the performance of functions. When it comes to the hard problem, the standard approach has nothing to say." (Chalmers 2010, 9)

"We have seen that there are systematic reasons why the usual methods of cognitive science and neuroscience fail to account for conscious experience. These are simply the wrong sort of methods. Nothing that they give to us can yield an explanation. To account for conscious experience, we need an *extra ingredient* in the explanation. This makes for a challenge to those who are serious about the hard problem of consciousness: what is your extra ingredient, and why should *that* account for conscious experience?" (Chalmers 2010, under point 5. "The Extra Ingredient", 13)

From the perspective of the SPP model it is difficult to understand that there is a gap. The SPP model provides proposals of how basic needs, higher needs and the library of associative-emotivational patterns and sensorimotor-emotivational patterns are developing for the sake of self-regulation, and of how inner experiences and evaluation and assessment processes are constantly ongoing in the brain in this context. If one follows these proposals, the consequence is that the major brain processes are always inevitably connected to emotivationally coloured experiences. Emotivational evaluation as a constantly ongoing inner experience is rather seen by the SPP model as one of the most important basic principles.

So the question of the *extra ingredient* that might be needed to obtain explanations of subject and experiences exists only outside of the SPP model, while it is inside of this model provided by the emotivational inbounds.

But it has to be considered that the SPP model maintains only a reduced view of the full amount of details. This is a sufficient method to explain some crucial interrelations on a hypothetical basis. But it cannot explain everything and the dilemmas which are described in Chalmers 2010 return immediately when this specific model has to be put aside for further exploration.

(4)

A specific question is asked in Part II "The science of consciousness", Chapter 2 "What is a neural correlate of consciousness?" (Chalmers 2010, 59ff.):

"The cornerstone of recent work in the neuroscience of consciousness has been the search for the 'neural correlate of consciousness.' This phrase is intended to refer to the neural system or systems primarily associated with conscious experience. The associated acronym is NCC. The hypothesis is that all of us have an NCC inside our

head. The project is to find out what the NCC is. In recent years there have been quite a few proposals about
the identity of the NCC." (Chalmers 2010, 59)
Chalmers follows up with a discussion of more detailed
questions and proposals.

Along with the SPP model it is proposed that the NCC is
provided by the neural correlate of the preparation mode of
the brain. These are the areas that can be attributed to the
attention assessment controller (AAC) and to all areas that
are potentially involved into the evaluation processes which
are constantly ongoing on control levels (4) and (5) of the
central nervous system.

Bieri, Peter: Analytische Philosophie des Geistes

The philosophical concepts of materialism, behaviourism,
functionalism, physicalism, reducibility, intentionality, mental states and representations, self-reference, self-awareness,
the subjective character of experience and of different kinds
of dualism, are discussed in the book. Contributions from
different authors are compiled together, providing complex
discussions on these matters.

A lot of crucial questions are asked, such as, for example: How can it be explained that mental phenomena evolve
from physical, chemical and biological processes occurring
in a system made up of molecules, cells and neurons? But
the results are poor. The discussions typically conclude with
statements like: "The current discussion on the question,
what can be concluded from phenomenal qualities for the
concepts of functionalism, did not lead to a clear result, but
seems to turn to the disadvantage of the functionalism. If

this tendency cannot be changed in the future, the functionalism as a common theory of the mind will be failed." (Bieri 2007, 51)

The book seems to make clear that the "analytical philosophy of mind" is nowadays nothing more than the maintenance of a huge lack of knowledge, accompanied by untenable philosophical speculations.

But the question of how mental phenomena and experiences might occur can easily be answered from the perspective of the SPP model. According to the SPP model, mental phenomena occur as inner experiences in terms of the direct and hypothetical projection of information to the evaluative focal plane, in which the brain plays the role of a medium to reflect and combine informational and emotivational patterns with no limit. For more details refer to the section entitled "Conscious experiences" and the preceding descriptions in Part 1.

There are at least three topics that are explained in the book which might correspond to the SPP model to a certain extent:

Chapter 5. "Hilary Putnam: The nature of mental states" in Bieri 2007, 123ff., covers the following statement:

> "All organisms who are able to suffer pain are probabilistic automata." (Bieri 2007, 128; original: Putnam 1975)

It is explained that this kind of automata bases decisions and behaviours on transition probabilities between (functional) neural states and between neural patterns etc., rather than on logical rules.

It is recommended that how these theses might correspond to the suitability probability based pattern selection process, which is proposed as a crucial principle by the SPP model, should be investigated further.

Chapter 12. "Thomas Nagel: What is it like to be a bat?" in Bieri 2007, 261ff., covers the following statement:
> "But in general an organism has conscious mental states then and only then, if it is in some way to be this organism – if it is in some way *for* this organism." (Bieri 2007, 262; original: Nagel 1974)

In the further texts it is emphasised that explanations of mental phenomena are not sufficient as long as the subjective character of experience is not included at the same time.

These theses should correspond to the basic statement of the SPP model that self-regulation or a purposeful component is a primary precondition for consciousness.

In the second part: "Intentionality, speech and thinking" in "Peter Bieri: Introduction" in Bieri 2007, 139ff., the concepts of "intentionality" are explained, as discussed recently, and it is mentioned that "mental representation" is a topic which is today the focus of the philosophy of the mind. These concepts refer to cognitive science, and the latter topic is particularly concerned with mental representations in terms of mental objects with semantic properties.

It is obvious that the concepts of intentionality and mental representations might be more detailed explanations of what is called direct and hypothetical projection in the section entitled "Conscious experiences" in the chapter about "The SPP model" in Part 1.

Psychology

Maslow, A. H.: Motivation and Personality

Some people say that Maslow's book from 1954 is obsolete (see Maslow 1987, first edition 1954). Some argue that the pyramid of needs is too schematic and inflexible, others argue that Maslow is no longer applicable to modern living conditions, where existential threats are widely excluded and basic needs do no longer shape the course of life. Both arguments are actually wrong.

Maslow did not primarily publish a pyramid of needs, but rather a motivation theory with emphasis on the "Dynamics of the Need Hierarchy" (Maslow 1987, 17). He gives also some examples of how needs can be categorized, but he warns at the same time that concrete categorisations can only be provisional and incomplete. However, a more crucial statement in his book from 1987 (1954) is rather the following:

> "It is quite true that humans live by bread alone – when there is no bread. But what happens to their desires when there is plenty of bread and when their bellies are chronically filled? [...]
>
> *At once other (and higher) needs emerge* and these, rather than physiological hungers, dominate the organism. And when these in turn are satisfied, again new (and still higher) needs emerge, and so on. This is what we mean by saying that the basic human needs are organized into a hierarchy of relative prepotency." (Maslow 1987, 17)

This characterises the actual message from Maslow, not the pyramid of needs, which he is famous for in this day and age. This means not that the pyramid of needs is useless, but it is important to understand that it is more a means to discuss the *dynamics* rather than a scheme which represents a final validity on its own.

The second argument – that we are today seldom confronted with existential threats and that our life is no longer shaped by basic needs – might be valid in one sense, but there are also lots of arguments, why this is not valid in other senses. Yes, modern life under affluent circumstances has a strong tendency to exclude basic experiences and to replace it by fashion, technology, science and fun. But real life is nevertheless characterised by basic needs because of several reasons. Few examples are just the following:

- Birth, death and disease can never be excluded.
- There are always places in our interconnected world where people suffer pain, poverty and deprivation. There are always also deplorable biographies in the best of all societies. All these incidents are relevant for each of us, even though one denies this.
- Mastering of basic needs occurs also as daily incident in the affluent society. This is kept away from consumers, but consumers might at the same time be professional workers who have to manage basic needs and tackle reality for consumers. This can be done in a professional way, widely supported by modern technologies, but there are always remaining difficulties and risks, which bring hardship to professional workers.
- The higher need levels are characterised by lower need levels and basic needs in two ways:
  1. The purpose of higher needs is adopted from lower needs. They get their vigour from basic needs via *emotivational dependency chains, emotivational stimulus modulation,* emotional and motivational components in the striving for rewards (in the reward system) or via other mechanisms. See also section "Higher needs" in chapter "The

SPP model" in Part 1.

2. Higher need levels are characterised to a great extent by representations of lower needs on each level above. Higher needs are often interspersed with components from lower need levels. See also the examples of "corresponding representations on each level above" in section "Dynamics of the need hierarchy" in chapter "The 4DI model" in Part 1.

Thus, basic needs and lower needs are the actual truth behind each more sophisticated demand. This is a rule which forces itself by the nature of the human nervous system and by *the character of consciousness* (to say it with Chalmers 2010). See also section "Fading consciousness in affluent contexts" in chapter "The 4DI model" in Part 1.

Kahneman, Daniel: Thinking, Fast and Slow

A central book of Kahneman 2012 is that the brain is equipped with two systems or two modes of thinking, respectively:

"Psychologists have been intensely interested for several decades in the two modes of thinking [...]. I adopt terms originally proposed by the psychologists Keith Stanovich and Richard West, and will refer to two systems in the mind, System 1 and System 2.

- *System 1* operates automatically and quickly, with little or no effort and no sense of voluntary control.
- *System 2* allocates attention to the effortful mental activities that demand it, including complex computations. The operations of System 2 are often associated with the subjective experience of agency, choice, and concentration." (Kahneman 2012,

"PART 1. TWO SYSTEMS", 20f.)
This correlates with the theses of the SPP model as follows:
- *System 1* refers to control levels (1)-(4) and to the unconscious part of the situational preparation processes, which perform always quick and highly efficient in relation to the current environment of the individual.
- *System 2* refers to control levels (4)-(5), the conscious part of the situational preparation processes and to the creative preparation processes in general.

It is a starting point in Kahneman 2012 that the human tends to see himself as a rational being. The main effort of the book is then spent on explaining that and why this is incorrect. The peculiarities of irrational decision making of humans, which are mostly produced by the strong influence of System 1, are the main topic over all parts of the book after part 1, which describes the two systems. Part 2 faces to "heuristics and biases", part 3 to "overconfidence", part 4 to typical behaviours of humans in making "choices" and part 5 to the "two selves" – the experiencing self and the remembering self – which deviate in their perceptions. In relation to part 4 of his book, Kahneman explains, for example:

> "The focus of part 4 is a conversation with the discipline of economics on the nature of decision making and on the assumption that economic agents are rational. This section of the book provides a current view, informed by the two-system model, of the key concepts of prospect theory, the model of choice that Amos and I published in 1979. Subsequent chapters address several ways human choices deviate from the rules of rationality. I deal with the unfortunate tendency to treat problems in isolation, and with framing effects, where decisions are shaped by inconsequential features of choice problems. These observations, which are readily explained by the

features of System 1, present a deep challenge to the rationality assumption favored in standard economics." (Kahneman 2012, 14)

The SPP-4DI model has an opposite approach. It sees the brain as a machine which is able to select neural patterns in highly efficient manner and to align this process to the requirement to act in the natural environment for the own benefit. It is postulated that this occurs as suitability probability evaluation process in a system which is described as *attention assessment controller*. It is clear that this cannot a priori result in rational decisions in the mathematical/statistical/economic sense. The brain and its evaluation and decision making apparatus follows many other maxims, which emerged from earlier stages of the evolution. Rational decisions are seen as possible, but only under special circumstances, namely when the 4D stress field can be managed in a good way (see chapter "The 4DI model" in Part 1 of this book). Hence, the human cannot be seen as a rational being per se, but rather as an intuitive being who might also be able to cultivate rational views under special circumstances with special efforts.

Kahneman describes among other things that we tend to make irrational decisions because of a "Loss Aversion" – see pages 283ff. in Kahneman 2012. It is explained that we dislike losses of any kind and that we are able to make unfavourable decisions, seen on a rational or mathematical point of view, because of this antipathy. This is seen by Kahneman as one of the many irrational potentials of humans.

From the perspective of the SPP-4DI model, the *loss aversion* is neither seen as disadvantageous nor as irrational. To lose an object or a piece of the familiar environment is always accompanied by the risk to lose a piece of behavioural effectiveness from the neural pattern library, which must be substituted again by more or less effortful training or exploration. That's why it is mostly better to prevent losses initially, although there are promising gains on the other side of the coin. I am sure that the human is able to overcome his *loss aversion*, but this is only recommendable if there is enough time to include also the potential neural cost into the balances by thorough reflection. For quick decisions it is always better – and also more rational – to prevent losses of any kind empathically.

Kahneman describes the human being as claiming for rationality, but as burdened with several biases and irrational intuitions. But I see this in a different way: There is only one real bias: to see the human as a rational being; the other biases and irrational seeming intuitions are rather natural peculiarities of human mind. Besides this should both approaches be consistent to each other.

Brain and computer

Commonalities and differences between brain and computer

There are many differences between brain and computer, but there are also commonalities on an abstract level.

Note: Here is the talk of the human brain and of the classic computer with von Neumann architecture or with a similar processor type.

Commonalities

1) Both brain and computer are **information processing engines**. They process information and control behaviour, based on external information, internal states, and stored data.

2) Both brain and computer are potentially also **engines which control states and behaviours of a larger system**. In this case, they do their work based on external information, internal states of the larger system, and stored data.

    Examples of *larger systems*, which can be controlled by brains are: biological organisms, mammals, humans. Examples of *larger systems*, which can be controlled by computers, are: machines, robotics, production lines, infrastructure facilities.

3) Both brain and computer **produce continuous sequences of operations**, at which subsequent operations are dependent on precedent operations, external information, internal states and stored data. The results of these sequences of operations are sequences of in-

formational patterns, which can be stored, sent to external systems or exploited as behaviours. Both brain and computer can interact with other systems by this way.

This can lead to various capabilities and performances for both brains (mammals, humans etc.) and computers (machines, production lines etc.), including solution of complex problems and goal-oriented behaviour over a long time.

4) Both brain and computer are **dedicated to an abstract purpose**.

Differences

1) The brain is a biological system. Information is represented in this system in the form of electrochemical excitations in nerve cords, at which the excitations have analogous values (firing rates) between zero (no excitation) and extremely high. These excitations are processed via networks of synapses in gradual manner. Data is stored persistently in the form of patterns of analogous values by establishing appropriate synaptic networks and by varying the degrees of excitation and inhibition at the involved synapses. Interactions are done via sensory organs (inbound), and muscles, limbs and glands (outbound).

The computer is a technical system. Information is represented in this system in the form of dual states, which are called Bits, and which can only have two states: true or false. This information is processed via arithmetic logic processors in digital manner. Data is stored persistently as digital patterns (Bits and Bytes) on storage devices (e.g. hard disks). Interactions are done via sensors and measurement devices (inbound), and motors, switches and any types of actuators (outbound).

2) A brain is typically utilised for control the behaviour of a biological organism (his owner, which might be a mammal or a human). Internal states of the system are bound into the brain via hypothalamus, endocrine system, pain sensation etc.

   A computer can be utilised for control the behaviour of a technical system (e.g. a machine, a production line). Internal states of the system are bound into the computer vial special sensor systems and interfaces, which are established for this purpose.

3) The selection of subsequent operations, based on precedent operations, external information, internal states and stored data, occurs in completely different manner for brains and computers. This is done by continuous pattern evaluation processes in neural networks in brains (see chapter "The SPP model" in Part 1), and it is done by binary decision logic and arithmetic operations in computers (see concepts "special purpose registers" – "instruction pointer" and "arithmetic logic unit"; see how conventional programming languages work).

4) The inherent (abstract) purpose of a brain is: thriving of its owner and species under the circumstances of evolutionary competition.

   A computer has no inherent purpose. It follows the purpose that has been dedicated to it by its inventor and user. So it follows the inherent purpose of brains, but only indirectly – as a tool.

The biggest open questions

The brain topic provides always much more open questions than answers. This book tries to discuss some possible answers on an abstract conceptual level, but the uncertainties, details which have yet to be clarified and remaining open questions are huge. It is not the intention to discuss these gaps in detail, but four of the questions will be mentioned here.

Question 1

A special question, which is extraordinary important, arises in relation to interactions between the basal ganglia loops (see chapter "Basal ganglia (BG) and frontal cortex") and the memory system (see section "The types of memory").

It is known that the basal ganglia circuits support parallel processing and synchronisation of the major part of the excitation processes in the brain. That excitation processes are controlled widely independently in parallel is known for skeletomotor control, oculomotor control, associative processes and limbic processes – see chapter "Basal ganglia (BG) and frontal cortex", section "The four parallel circuits through the basal ganglia (BG)". But what cannot be clarified here (in this book) is the question: **How are the excitations, which are controlled independently in parallel by the four BG circuits, divided from one another in the involved cortical areas?** There are always also pathways and interdependencies between the four segregated systems, and it is not very clear, how it is achieved that the systems are able to act independently on the first hand, to interact on the second hand and that this does not lead to total confusion and chaotic flow of excitation on the third hand.

Regarding memory, it is known which parts of the associative areas are to be attributed to the declarative side or rather to the procedural side. But there must be an explicit border, because memory reinstatement and associative processes on the one hand and sensorimotor activity sequences on the other hand can be independently controlled in parallel by the circuits in the basal ganglia, what might only be possible if excitation does not flow unhindered to and fro between both sides. It is likely that this question can be answered on the basis of knowledge about the detailed domain specific functions of each concrete associative area or of each other brain area. But this goes beyond the intended scope of this book.

Question 2

It is clear that somatosensory perception, sense of touch and other senses – as vision, hearing, taste, smell etc. – are incorporated as informational criteria into decision making processes. This occurs probably partially via cascades of perceptual areas and modules and their final pathways to the areas that are responsible for decision making, partially via involvement of perception into situational sensorimotor control processes, which are the actual target of the decision making processes, and partially via microrepresentations of sensorimotor patterns in the declarative memory (as remembered perception incidents).

But what is not clear, or – more precisely – what is not made clear in this book is the question: **How is perception processing precisely bound into the decision making processes?** Decision making occurs particularly in the basal ganglia (as affective decision making), the prefrontal cortex (as value-guided decision making) and in the anterior cingulate cortex (as motor decision making). But it is an open field, which is not discussed in this book, how excitation patterns which represent perception (in the informational sense) are precisely transferred from the sensory organs via the perceptual areas to the decision making instances of the brain. This is a matter which has still to be investigated.

Question 3

It is clear that there must be an emotivational system, which involves signals from the amygdala (representing emotions), from the autonomic interface (consisting of hypothalamus, endocrine system etc.; representing homeostatic regulation and balancing between sympathetic and parasympathetic nervous system), from pain sensation (via thalamus) and from the reward system, and which forces self-regulation and optimisation all over the brain by primarily taking effect in the decision making system (basal ganglia, prefrontal cortex, anterior cingulate cortex) and in the declarative memory (entorhinal cortex, hippocampus etc.).

But a collection of remaining questions is: **How does the emotivational system work in detail?** What are the concrete pathways, which represent this system? How is it achieved that all types of emotivational signals are combined in a manner so that they build a universal assessment and evaluation system, which gives meaning to the brain processes and to the complete organism?

Question 4

Questions 2 and 3 are closely linked with the explanation of consciousness and conscious experience. Section "Conscious experiences" in chapter "The SPP model" in Part 1 explains conscious experience as projection of informational excitation patterns to the evaluative focal plane or to the signals of the emotivational system or to the *attention assess-*

ment controller, respectively. This involves at least the following brain areas: basal ganglia (BG), prefrontal cortex (PFC) and anterior cingulate cortex (ACC). But this is only an abstract book, which needs further investigation in order to achieve profound knowledge in this regard.

The book from section "Conscious experiences" might be a good starting point for the systematic research of consciousness. But a collection of remaining questions is: **How is consciousness realised in detail in the brain?** Where are the locations of the evaluative focal plane in detail, i.e. in which concrete modules and circuits meet informational and emotivational signals each other and how are they resolved in relation to each other? How can projection at the evaluative focal plane be measured and investigated in empirical manner? How can the concrete role of the posterior cingulate cortex (PCC) or of the networks around the complete cingulate cortex be defined in this connection and in general?

References

Alkire, M. T. (2008). General anesthesia and consciousness. In S. Laureys (Ed.), The neurology of consciousness: Cognitive neuroscience and neuropathology: Amsterdam: Elsevier.

Alkire, M. T., Haier, R. J. and Fallon, J. H. (2000). Toward a unified theory of narcosis: Brain imaging evidence for a thalamocortical switch as the neurophysiological basis of an anesthetic-induced unconsciousness. Conscious. Cogn., 9(3), 370– 386.

Alkire, M. T. and Miller, J. (2005). General anesthesia and the neural correlates of consciousness. Prog. Brain Res., 150, 229–244.

Bateson G. (1972). Steps to an ecology of mind. Chicago, IL, University of Chicago Press.

Bieri, Peter (Hrsg.) (2007). Analytische Philosophie des Geistes. 4. Auflage. Beltz Verlag, Weinheim und Basel.

Block, Ned (2009). Comparing the Major Theories of Consciousness. New York University, New York, New York (online).

Chalmers, D. (1996). The conscious mind: In search of a fundamental theory. Oxford, UK: Oxford University Press.

Chalmers, David J. (2010). The Character of Consciousness. Oxford University Press, Inc.

Chan, Shiao-hui, Ryan, Lee and Bever, Thomas G. (2013). Role of the striatum in language: Syntactic and conceptual sequencing. Brain & Language 125 (2013), 283–294. Elsevier.

Crone, Eveline A. (2014). Cognitive Control and Affective Decision Making in Childhood and Adolescence. In Gazzanega and Mangun 2014, 23ff.

Davachi, Lila and Preston, Alison (2014). The Medial Temporal Lobe and Memory. In Gazzanega and Mangun 2014,

539ff.

Douglas, Rodney J. and Martin, Kevan A. C. (2014). Organizing Principles of Cortical Circuits and Their Function. In Gazzanega and Mangun 2014, 79ff.

Eagleman, David (2007). 10 Unsolved Mysteries Of The Brain. Discover Magazine (online).

Gazzanega, Michael S. and Mangun, George, R. (Editors in Chief) (2014). The Cognitive Neurosciences. Fifth Edition. A Bradford Book. The MIT Press, Massachusetts Institute of Technology, Cambridge, Massachusetts.

Glimcher, Paul (2014). The Emerging Standard Model of the Human Decision-Making Apparatus. In Gazzanega and Mangun 2014, 683ff.

Harari, Yuval Noah (2014). Sapiens. A brief history of humankind. Harvill Secker.

Huettel, Scott A. (2014). Neuroeconomics. In Gazzanega and Mangun 2014, 1063ff.

Kahneman, Daniel (2012). Thinking, Fast and Slow. Penguin Books Ltd.

Laureys, S. (2005). The neural correlate of (un)awareness: Lessons from the vegetative state. Trends Cogn. Sci., 9(2), 556–559.

Levine, J. (1983). Materialism and qualia: The explanatory gap. Pac. Philos. Q., 64, 354–361.

McGinn, C. (1991). The problem of consciousness. Oxford, UK: Oxford University Press.

Maslow, Abraham H. (1987). Motivation and Personality. Third Edition. Addison Wesley Longman, Inc.

Nagel, T. (1974). What is it like to be a bat? Philos. Rev., 83(4), 435–450.

Ochsner, Kevin N. (2014). What is the Role of Control in Emotional Life? In Gazzanega and Mangun 2014, 719ff.

Osgood, C. E., Suci. G. J. and Tannenbaum, P. H. (1957). The measurement of meaning. Urbana: University of Illinois Press.

Popper, Karl R.; Eccles, John C. (2006). The Self and Its Brain. An Argument for Interactionism. Routledge, London and New York.

Putnam, Hilary (1975). The nature of mental states. In Mind, Language and Reality, Cambridge 1975, 429-40.

Ranganath, Charan, Hasselmo, Michael E. and Stern, Chantal E. (2014). Short-Term Memory: Neural Mechanisms, Brain Systems, and Cognitive Processes. In Gazzanega and Mangun 2014, 527ff.

Rushworth, Matthew F. S., Chau, B- K. H., Schüffelgen, U., Neubert, F.-X. and Kolling N. (2014). Choice Values: The Frontal Cortex and Decision Making. In Gazzanega and Mangun 2014, 501ff.

Schneider, G. and Kochs, E. F. (2007). The search for structures and mechanisms controlling anesthesia-induced unconsciousness. Anesthesiology, 107(2), 195–198.

Sherrington, C. (1908). On the reciprocal innervation of antagonistic muscles. Thirteenth note. On the antagonism between reflex inhibition and reflex excitation. *Proc R Soc Lond B Biol Sci, 80b*, 565-578 (reprinted in *Folia neurobiol*, 1908, 1, 365).

Singer, Wolf (2004). Selbsterfahrung und Neurobiologische Fremdbeschreibung. Zwei konfliktträchtige Erkenntnisquellen. In: Deutsche Zeitschrift für Philosophie 2 (2004), 235-255.

Tononi, G. (2012). Integrated information theory of consciousness: an updated account. Archives Italiennes de Biologie, 150, 290–326 (online).

Tononi, G. and Koch, C. (2008). The neural correlates of consciousness: An update. Ann. NY Acad. Sci., 1124, 239–261.

Turner, Robert S. and Pasquereau, Benjamin (2014). Basal Ganglia Function. In Gazzanega and Mangun 2014, 435ff.

Velly, L., Rey, M., Bruder, N., Gouvitsos, F., Witjas, T., Regis, J., et al. (2007). Differential dynamic of action on cortical and subcortical structures of anesthetic agents during induction of anesthesia. Anesthesiology, 107(2), 202–212.

Wagner, Dylan D. and Heatherton, Todd F. (2014). Self-Regulation and Its Failures. In Gazzanega and Mangun 2014, 709ff.